THE
QUARRY FOX
AND OTHER CRITTERS
OF THE WILD CATSKILLS

THE

QUARRY FOX

AND OTHER CRITTERS
OF THE WILD CATSKILLS

LESLIE T. SHARPE

THE OVERLOOK PRESS
NEW YORK, NY

This edition first published in hardcover in the United States in 2017 by
The Overlook Press, Peter Mayer Publishers, Inc.

NEW YORK
141 Wooster Street
New York, NY 10012
www.overlookpress.com
For bulk and special sales, please contact sales@overlookny.com,
or write us at the above address.

Cataloging-in-Publication Data is available from the Library of Congress

Book design and typeformatting by Bernard Schleifer
Manufactured in the United States of America
FIRST EDITION
ISBN 978-1-4683-1247-8
2 4 6 8 10 9 7 5 3 1

For my parents, Hannah and Arthur Sharpe,
who loved all critters.

With special thanks to Dr. Julie Reece, my greatest teacher,
who nurtured every plant, critter, and person she met.

In Wildness is the preservation of the world.
—HENRY DAVID THOREAU

CONTENTS

INTRODUCTION

The Courage of Crustaceans

(For H. D. Thoreau)

EVERY SPRING, I CLEAR LITTLE PORCUPINE CREEK. I HEAD
into the "Cut"—a natural opening in the woods that edge the
meadows that surround my cabin in the Great Western
Catskills of upstate New York—armed with a rake and a machete that
can do battle with the thorny, vindictive wild rose that has invaded these
hills. When I first explored the land on this mountainside, the creek was
a bog, already overgrown in May with dark-green leafy watercress eager
to leech out the stream that trickled tenuously beneath the tangle of
sticks, dead leaves, and rocks that crowned it. But still visible beneath
last winter's leavings was the outline of a narrow channel that would
split in two, then become one again, meandering down the mountain
until it curved into deeper woods and disappeared.

Its source, I quickly discovered by tracing the channel a short way
uphill, was a spring that emerged out of a small, cave-like opening in
the mountain. Even though the stream was quickly subsumed into the
oozy bog, I could still hear it running. I *had* to free that water! I knew it
would be a benefit to the animals and birds, too, that lived in these
woods, and I longed to sit in the seat that Nature had carved into the
large bluestone boulder (which I had already dubbed "Reading Rock"),
listening to the creek as it rushed around it. Turning the bog into a flowing stream (Little Porcupine Creek is named for a young porcupine I
startled while out on a walk, which clambered up a white pine tree and

which I left, allowing it to flee with dignity) was hard work, especially that first spring—it took several days to weed, hack away the thick, stubborn undergrowth, and excavate the bog of rocks. But the sight of the water running free again, hearing that music of the mountain in concert with the wind in the aspens and the melancholy, flute-like song of the hermit thrush at dusk, returned for summer, was a reward that cannot be calculated.

Clearing the creek every spring, in May, of its mantle of detritus is a ritual as well as a necessity—Nature quickly overcomes here, as the crumbling rocks, the foundation of what was probably once a sugar house, judging from the rusted tin sap bucket I found in a nearby grove of maples, attests to. This work, strenuous and slow, binds me to the mountain, gives me an intimate knowledge of its rocks, its dirt, connects me as well to its previous tenets: the Lenni Lenape Indians, who lived along the banks of the Delaware River and whose sandstone arrowheads I have found in the creek, and generations of farmers too, who struggled to cultivate the recalcitrant red clay soil rife with stone before abandoning it to sheep and cattle for foraging. The banks of the stream, especially after rain, also reveal the tracks of the same animals my forebears here knew—critters great and small, from black bear to white-footed mouse, and now that the waters flow freely, a ruby-throated hummingbird family has taken up residence on a white pine bough overhanging the creek, building its teacup nest there every spring.

I always wear sturdy leather gloves—raking rocks readily causes blisters—and Wellington boots because water won't penetrate them (though heavy wool socks are a must to resist the chill), when working in the stream. This May, as I raked the rocks to one side, knowing full well that spring rains would topple them back into the streambed, the water became turbid and impatient, I reached down and moved the rocks by hand, instead of waiting for the water to clear.

I am always careful picking up a rock on land. It can house beneath it snakes as benign as a garter snake, with its three yellowish hor-

izontal stripes (its "garters," according to
my favorite version of how this critter got
its name), which can still startle, or as
dangerous as a venomous timber rattler,
with its characteristic spade-shaped head,
especially in the Catskill high hills. But I
didn't expect to disturb a denizen living
in the stream—a crayfish, I realized,
looking exactly like a miniature lobster,
its relative, a tiny crustacean about three
inches long.

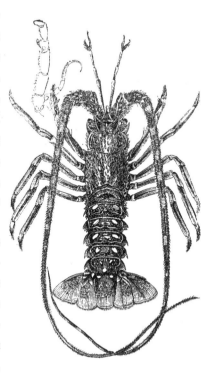

Delighted, I reached down to
scoop up the crayfish for a brief examina-
tion (crayfish breathe through delicate
gills, which allow it to extract oxygen
from water), but it seized my finger, fortunately gloved, with one of its two front
claws and locked on—dangling in the air.

As I cupped my other hand beneath the crayfish so it wouldn't fall
back onto the rocks, I had to laugh. I have always admired the courage
of crustaceans. Then, I took the opportunity, however briefly, to examine
it, my curiosity kicking in. I counted ten legs, including its front two,
which boasted large claws that crayfish use to defend themselves from
enemies as well as to attack prey and even to move around small objects
such as stones. Like its crustacean cousins, lobsters and crabs, which feature a tough, protective exoskeleton that helps thwart predators (and
which requires molting as the critter grows), the crayfish had a jointed
head and thorax (or midsection), and a segmented body. Its dark-brown
color, I noted, was another defense, allowing the crayfish to blend in perfectly with the muddy stream bottom. Feisty and resourceful, crayfish
are scavengers as well as predators, seeking out insects and frog eggs,
cleansing the creek of plant and animal matter. In the cycle of life, it is

also prey to those other rock lurkers, snakes, and a variety of mammals, especially here, raccoons and opossum. I was pleased to meet this crayfish—its presence confirmed that the creek was a healthy ecosystem. Discerning as well as cantankerous, crayfish will not tenant polluted streams or other freshwater bodies.

Holding it in my gloved hand, I looked especially at its face—its pointy snout, its two long antennae, whiskery feelers, and its tiny black eyes perched atop their movable stalks. What, I wondered, did the crayfish make of me? And what, I wondered, would it be like, to call this creek a kingdom, the rock, which I had so rudely disrupted, a castle? Finally, I put the crayfish, still grasping my finger with its resolute claw, back into the creek, where it abruptly released me and flipped its tail in triumph, propelling itself backwards, quickly disappearing into the roiled creek water.

I had held the crayfish for mere seconds but in that time I had glimpsed another world—the world of a tiny creature perfectly adapted to its environment, whose crustacean cousins, years before, had afforded me my first lessons as a naturalist. When I was a child, spending summers at the seashore, one of my favorite activities was "crabbing." But the crabs that we caught in the trap my father dropped from the dock on the bay were not captives. We would bait the metal trap with its four collapsible sides with a fish head. Then I would wait, breathlessly, pulling up the trap after seconds to find nothing. My father would laugh and urge patience. If I were *too* patient, waited too long, the trap would be empty—the clever crabs and eels, too, had stripped away the bait. But if I were lucky, my patience would be rewarded by the prize: the Atlantic blue crab, those "beautiful swimmers." The captured crab, which could measure nine inches across its shell, or carapace, would scuttle out of its rusted cage, waving its sapphire-tinted claws—such a brilliant hue against its duller olive green or mottled brown shell—which give this otherwise modest crustacean its name, then clamber sideways on its three pairs of walking legs until it scooted off the dock into the bay. On one occasion, our dachshund Rudolph, barking furiously, decided to challenge this strange armored invader. The blue crab, unimpressed—and unintimidated—held its ground, then tweaked the charging dog's nose with one of its powerful pincers to much yelping (lesson learned!), before flying off the dock, brandishing its claws like a gunslinger wielding six-shooters. Another day, while swimming in the bay, I felt a sharp pinch on my little toe. Looking back through the goggles I was so proud of, I saw a blue crab swimming down into the depths, its two back legs propelling it like paddles—the enterprising blue had bitten off more than it could chew, literally. That nip left a brown bruise on my toe for what seemed like forever, a souvenir of summer that I treasured, proof that I, too, was a creature of the sea.

We would also pull up in the metal trap spider crabs, ugly and gangly, especially compared to the elegant, lithe blue, with long "spidery" legs, its shell often festooned with seaweed and barnacles. But the spider crab, awkward and bumbling on the dock's wooden boards, gave me my first insight into adaptation as a means of survival—its muddy brown coloration and its shell's rough, scabby surface that caught bits of the bay bottom allowed it to blend into its habitat as both prey and predator.

It was the unlikely crustacean, whose foreignness fascinated me and whose courage captured my child's heart, which sparked my lifelong passion for the natural world. I even remember the exact moment I became a naturalist. I was no more than seven, digging in the sand at the seashore, engaged in another favorite activity, building drip castles. First, I excavated sand to make a small slurry near the shoreline, then "dripped" wet sand from my fist to create an assemblage of Gaudiesque turrets. Once, starting a new slurry, I was startled by an apparition that appeared, swimming in the miniature saltwater basin. It was small, no bigger than a thumb, the color of wet gray sand, with pointy legs but no pincers. Replacing the odd-looking critter, its legs wriggling, in the slurry, I was delighted to see it burrow backwards into the sand and disappear. My discovery of the mole crab, or "sand crab," as we called it, a tiny crustacean that makes its living at the sea's edge where waves crash against the shore, using its feathery antennae to catch minute organisms for food, was my first intimation that the world was both larger and smaller than I knew—a world mostly unseen by humans and indifferent to them, a world that held as many secrets as it did surprising and secretive critters, such as the elusive, ghostlike mole crab.

The curiosity about critters the crustaceans awakened in me as a child has never left me, and the patience my father urged while "crabbing" for the illustrious Atlantic blue has served me well when watching wildlife. Curiosity and patience are requisites for the naturalist, someone who observes the natural world, its animals, and plants in their own setting. Through dedicated observation that can span days, seasons, and even years, the naturalist learns to discern patterns and their meaning in animal behavior and also appearance. The male American robin, with its darker head and bright orange breast, precedes the females north in spring to scope out the best nesting sites with which to woo a mate. (In this case, the early bird gets the girl as well as the worm.) The female, once she has assessed a suitor's property as well as singing voice, having followed the males "home" from several days to two weeks later, will also have the benefit of mud, as opposed to the frozen earth, with which to build her nest of twigs and dry leaves. Monarch butterfly caterpillars only eat milkweed leaves and monarchs also need milkweed plants to lay their eggs, whose noxious chemicals render this insect so bitter-tasting that birds shun it. In a cunning adaptation, the viceroy butterfly, which, though not toxic, mimics the monarch's striking orange and black coloration, its distinctive patterning, and as a result, is largely avoided by predators.

Another necessary attribute of the naturalist is the ability to be amazed, which is how I feel when I hold a bluestone rock in my hand. It may seem an unlikely object of awe, but bluestone offers us a window into a time when much of New York—from the Hudson River to Lake Erie—was covered by a shallow, tropical sea. In this "Age of Fishes," sharks and other cartilaginous fish, such as rays, and the first bony fish abounded. The sea was fueled by swift-flowing rivers fed by torrential streams cascading down the sides of the Acadian Mountains, which towered over today's Hudson Valley, taller than the Rockies, long lost to time and erosion. The Acadians' rocky runoff was also carried into the warm sea, where deposits of mineral grains would eventually be galvanized into the bluestone that forms the core of the Catskill Mountains—an evenly layered sandstone that splits easily into smooth, thin slabs, which still make it a prized building material. Even today, it is possible to see this ancient sea and the racing rivers that nurtured it at work in ripple marks left on bluestone slabs, and in some places in the Catskills, the fossilized remains of clam burrows, visible in sandstone rock, attests to a time when these mountains were a primitive sea. My encounter with the diminutive crayfish, the "freshwater lobster," whose own crustacean forebears date back to the Paleozoic Era, had made me pause in the midst of clearing Little Porcupine Creek to wonder at the ancientness of these hills and to hear the echo of an ocean in the thunder of a spring day.

Perhaps that first connection I felt with crustaceans was the simple fact that they were little and so was I; if they could be brave, then I could, too. But it is, I believe, that connection—a very powerful and deeply felt connection to the natural world, as well as an appreciation of its beauty and a sense of commonality with its creatures—that marks the true naturalist, whether a wildlife biologist or a backyard birder, and differentiates him or her from the "pure" scientist conducting lab experiments and attempting to draw conclusions from data alone.

The naturalist's most essential tools are his or her senses. What you see—a young buck bounding away, suddenly aware of your presence, its upraised tail a furled flag of white. What you hear—the explosive whoosh of an American woodcock as it takes flight, flushed from the leaf litter of the forest floor. What you smell—the pungent fragrance of skunk, emitted to ward off a predator. (In winter, that sharp, acrid odor could also indicate a courting dog fox, signaling his amorous intent to a nearby vixen.) What you touch—the stickiness of sap rising in the trunk of a sugar maple in spring. If you wait, insects may appear, attracted by the sweetness—wait a little longer, and a yellow-bellied sapsucker alights on the tree—a woodpecker that eats both the sap and insects that find it so alluring. The next day, if you return, you will find a series of holes the sapsucker has drilled into the bark, in search of both delicacies. Running your hand over these small perforations, you discover their carefully worked pattern and are reminded of Nature's larger pattern here—how the tree and its sap, the insects and the woodpecker, are all connected.

Being a true naturalist also means being objective. On the mountain I try (and at times even struggle) to balance my passion for the black bears, bluebirds, and bobcats, among all sorts of critters, with the dispassion necessary to record the drama of their everyday lives with objectivity. I am the first to admit that I sometimes break my own first naturalist's rule—"never fall in love with wild animals"—not only to avoid sentimentality in writing about them but to escape the heartbreak of discovering a cottontail rabbit, the one with the nip in its ear, companion of my gardening days, dead in the driveway where a hunting red-tail hawk has dropped it.

Nature, whose only mandate is species survival, does not play favorites and has no regard for our own particular affections. Many more animals are lost than survive, which is why birds and small mammals, especially "prey" such as Eastern cottontails, reproduce so prolifically. For critters, the struggle to survive, simply to eat, is at the center of existence. A chipmunk, which yesterday marauded a robin's nest, stealing its sky-blue eggs, now intent on black-oil sunflower seeds scattered on the ground by black-capped chickadees busy at a backyard feeder, risks everything bounding across open land from its hiding place in the rocks at the forest edge. That shrewd hunter of the skies, the goshawk, stationed all morning atop the stately Norway Spruce, has been waiting and watching for just such an opportunity . . .

The cycle of life, of which death is such an integral and necessary part, is ever-changing Nature's one true constant. It rules everything in the wild Catskills and is immune to our most precious human precepts, perhaps especially fairness. Death, arbitrary and common, has not become easier for me to bear, though I understand its necessity, even in strengthening species. But it *has* sharpened my appreciation of life, deepening the sense of joy I feel when it triumphs. For every sorrow—three bluebird nestlings who perished in their nesting box during an unusually cold and wet spring—there are many successes: five bluebird siblings who fledged the next summer in the same box, returning together in October to rest in a barberry bush just before flying south, almost as if saying goodbye. That is one lesson I have learned living in the Catskills: that life will outlast death, just as nature, if given a chance, will always prevail.

I have also come to comprehend the meaning of the oft-invoked phrase "web of life." Everything in nature—even weeds—has a place and a purpose. The dandelion, that cheerful scourge of suburban front yards, is in fact a favorite of honeybees, now in decline, sustaining these crucial pollinators in spring until other wildflowers bloom. Another gardener's bane, the spiky, intimidating bull thistle, is key to the life cycle of goldfinches, which are dependent on its soft, feathery down for nesting material and its seeds to make a milky cereal for their young. And without the milkweed plant—a favorite of children for its pods bursting with silken strands, "parachutes" that spin in the wind, carrying seeds far and wide—which is rapidly disappearing from cleared land, the majestic monarch butterfly's very survival is in peril.

Such connections, fragile yet enduring, intricate yet simple, fill me with wonder. There is nothing so sublime that is so readily accessible as the natural world. To experience it, one only has to look, to train oneself to truly see, and perhaps most important, to allow oneself to be amazed.

THE
QUARRY FOX
AND OTHER CRITTERS
OF THE WILD CATSKILLS

CHAPTER ONE

Waiting for Spring

(For John Burroughs)

TODAY IS THE FIRST DAY OF SPRING. THE SKY IS A DARKENING gray, and all day the temperature has hovered around freezing. Gray is the color we do best in the Great Western Catskills of upstate New York, even in summer, when smoky wisps cling to the mountains after a thunderstorm like lingering spirits, which the Indians believed they were, and especially in early morning, when mists rise from the West Branch of the Delaware River, enveloping my house in a milky white cloud so opaque I can barely make out the steps that lead from the deck to the dandelion-strewn lawn. Here, we have perfected gray in all its gradations—we have grays for every season, every time of day. We have grays in every shade and texture.

On this first day of spring, it is the texture of gray that tells me it is going to snow—the clouds are thick, full, palpable to an imagined touch, so low in the sky that the prop-propeller plane taking off from the cow pasture in the valley at the foot of the mountain is quickly enveloped, lost from sight. As I stand at the window, looking out at a world still in monochrome—the barren brown meadows broken only by swathes of stubborn snow, the bare trees, their dark branches exposed, those caught out in the open like the Japanese maple I foolishly planted before I understood the elements here, gnarled silhouettes leaning away from the incessant wind—the only spikes of color are provided by the evergreens on the surrounding mountains, hemlocks, red and white pine, eastern red cedar and spruce and balsam fir, their varying hues the same at a distance, such a respite from their companions, the stripped naked bodies of deciduous trees. Then it starts to snow, the sooty clouds giving way to a cascade of frozen cinders.

But the squall, which is fierce, causing a virtual whiteout, dissipates as abruptly as it began. A white mist forms over the meadows, and as I watch, a flock of American robins returns to hunt for worms, easily finding the soft spots among the still-frozen patches of earth. I first saw the robins ten days ago, when an early spring storm dumped nearly two feet of snow on Lazy Hawk Mountain, as I call it, until the next caretaker of this land names my Catskill foothill anew. It was then, on March 10th, that I turned to the calendar for reassurance to see that spring—the "Vernal Equinox," that elegant, literary-sounding phrase—was just ten days away. That knowledge is what makes such a late-season snow, if not sweet, then easier to bear. The sun is so much stronger now, the snow will melt so much faster, despite winter's defiance. Indeed, that morning as I stepped out on the back deck to shovel to a gloriously sunny day, the sky so clearly and correctly blue without the subtleties of our usual grays, I heard the first true sound of spring—the quick running of small water. The steady *drip-drip-drip* of icicles that had adorned the roof for months signaled the release, however reluctant, of snow everywhere, and the beginning of a season within spring for those of us who live in the North Country: the mud season, which will last into June.

As I struggled to shovel my way out to the bird feeders on the edge of the western woods that morning of the Big Snow, I was met by a band of black-capped chickadees, the hardy little birds that overwinter here, clever, congenial, and curious—always the first to find feeders and alert other birds to them—whose friendly *dee-dee-dee* sparked with excitement at the sight of the bucket of black-oil sunflower seeds I had parked in the cleared path behind me. The temperature outside, I noticed when I left the house at 9:00 AM, was already 42 degrees, and it struck me how warm the sun felt on my face. If this were December or January, even such a windless day would feel so much colder, the bright sun a mere ornament in the sky. But spring's gift is the return of light, the lengthening of days that trigger the migrations north of our summer birds and, in March, the courtship finery of the male wild turkey, whose eager hormones, stimulated by this gift of light, begin to paint his head and neck with handsome colors—bold blues and vibrant reds—and coat his feathers with the shimmering iridescence that "Tom's" female consorts will find so alluring. Light is life: it's as simple—and as old—as that. And in the fall, it is the loss of light, its relentless diminishment, that triggers the migrations south, as seasonal residents leave to find warmer days, especially warblers and other insectivores, whose prey will die with the coming cold, as well as those who sip nectar, their flowers inevitably falling to the first frost, such as the delicate but doughty ruby-throated hummingbird, whose journey will begin at the close of August, signaling, for me, summer's true end.

The Big Snow was heavy and wet, as is characteristic of warm spring snows, and it also held a secret. As I shoveled the deep snow to one side, creating a pristine wall of perfect white, I was startled to see striations of blue in the snow bank. It was an eerie, evocative blue, a ghostly blue, the blue of ancient glaciers.

"Blue snow," I said out loud, in awe, as if I had just made this discovery. The secret of that snow that I was suddenly privy to has, I would learn, an explanation in science. But the soothing, right-sounding explanations of science could not quantify the wonder I felt at seeing blue snow for the first time, how it lifted my spirits, reminding me that in nature, there are always mysteries, even in an act as mundane as shoveling snow.

That heavy, wet snow that made for the crevasses of blue in the snow bank also made my journey to the feeders—a usually short distance from where I began at the dirt driveway's edge—interminable, and I had to caution myself to "lift with your legs, not with your back" as I shoveled. I was escorted every step of the way by the chickadees, and when I finally arrived at the feeders, their *dee-dee-dee* seemed to ricochet around me. They are as fearless as they are friendly—one perched on my right shoulder as another landed on the lip of the bucket, sampling the sunflower seeds that would soon be theirs. High above, in the branches of the Quaking Aspen trees that shelter the feeders—in the tallest of these straight, sturdy but elegant trees, their white bark luminous, as if reflecting the deep snow—crows, their slick black feathers gleaming in the morning sun, were watching me warily. They knew I would leave a special treat for them—roasted, unsalted peanuts in the shell—yet they did not greet me as the chickadees did, their opposites in size and temperament. They always wait until I have withdrawn into the house before they descend to pluck the peanuts—holding as many as three in their strong bill at the same time—from the onslaught of gray squirrels.

Crows are the smartest of birds, along with their larger cousins the ravens, residents of even higher elevations and fellow members of the corvid family (which also includes rooks, jackdaws, jays and magpies, clever critters all). Tool use, once thought to separate humans from other animals and mark us as superior—a supposition Jane Goodall dashed when, in 1960, she observed chimps at Tanzania's Gombe Stream National Park inserting small twigs into the ground to fish out insects for food—is also a skill of crows. They have been seen using a twig—held in the bill—to pry bark off of trees to get at the insects hiding beneath (which I have witnessed) and, like the chimps, have been known to insert a twig into a hole in trees to extract beetle larvae (hence, the crows' sometime nickname, "feathered apes"). I have also observed crows routinely dipping a piece of stale bread into a puddle to soften it, delighted by their cleverness. They have complex family structures, and within a flock, can work together for the common good.

A colony of crows has tenanted the white pines and hemlocks on Lazy Hawk Ridge, high above my house, since I have lived here, and they are my constant, noisy companions. The "crow police," as I call them, rout red-tailed hawks from the skies and buzz off blue jays, especially during nesting season, and once even descended to scold a red fox as it waited patiently for a white-footed mouse or meadow vole to emerge from the wildflower garden directly outside my bedroom window at dawn, waking me in the process and thwarting the fox, which eventually fled without breakfast, tired of being reprimanded. Again, this "mobbing" behavior is an example of the crow's ability—and that of many other birds, which also practice this strategy to repel predators, including larger birds such as the crows themselves—to work cooperatively. (I have also observed such inter-species cooperation—the unlikely trio of a male cardinal, red-winged blackbird, and Baltimore oriole banding together to chase a crow away from their nests. The crow took off and landed in a tree some distance away, which did not assuage the other birds. They followed the crow, landing on the branch above it and screaming at it until the browbeaten corvid finally retreated.) Whenever the crows caw, I stop whatever I am doing and watch for them, curious to see what they are up to. And their raucous calling, though perhaps annoying to some, is for me a companionable comfort, especially in winter when the mountain is silent and so few birds sing or even seem to fly. There are always the crows, hopping about comically in the snow, jet-black against the white like spilled ink on a pristine page, pilfering a peanut or two or three from an irate gray squirrel and bragging about it noisily to everyone on the mountain.

Before I filled the feeders with sunflower seed and replenished the "hot pepper" suet (to deter the squirrels, which quickly learn NOT to gobble these cakes—birds have far fewer taste buds than mammals and are unaffected by the spicy suet) in those feeders, I dug out a circle of snow from underneath them to spread seed for the ground-feeding birds—dark-eyed juncos (once, in Spring, when I ventured too close to her nest, the female performed a "distraction display," feigning an injured wing by dragging it along the ground to lure me away from her nearby young); my Northern cardinal pair, his regal red a jolt against the white landscape, his demure wife, her colors muted, soft brown accentuated by a quieter red; mourning doves; blue jays, too, boisterous busybodies who will challenge the crows and the squirrels for peanuts; tree sparrows; tufted titmice; white-breasted nuthatches; downy and hairy woodpeckers, all of whom I knew were hiding deeper in the woods, so hungry in this snow that would cover the ground and any food for days. Winter can be reluctant to release us here in the Catskills, can sometimes seem spiteful and vindictive, sending us snows into April, so resentful of the coming of light and color.

Having dug out the area well enough to allow the birds to land without sinking into the snow, I noticed a small round hole, directly below one of the seed feeders, but thought nothing of it. As I filled the feeders, some seed spilled over and fell directly into the hole, more scattered around it. As if on cue, a tiny head popped out, surveyed the scene—including me—and abruptly disappeared back into the hole. I stood very still. Unable to resist the shower of sunflower seeds, the head reemerged—a meadow vole, I realized. It disappeared once again at the sight of me, and I had to laugh. This "country mouse," with its round ears (the model for Disney's Mickey, perhaps?), stubby tail, and stout little body had moved into this hole directly below the feeder, manna from heaven delivered into its hideaway. I shook some seed into its hole as a reward. Cleverness counts, perhaps especially in small rodents, who depend on a cache of nuts and seeds to last through Winter to limit foraging expeditions, preferably under cover of snow, which is porous and allows them to breathe, providing warmth and some measure of protection from any number of predators living on the mountain: a red-tailed hawk; an elusive bobcat; a gray fox and red fox, too; a barred owl, hunting by day as well as at night; a lucky coyote; even a black bear, having woken up hungry on a mild winter morning.

I picked up the bucket, which felt weightless without the seed, took the shovel in my other hand as the chickadees alighted on the feeders, leading the way for the shyer birds, already starting to emerge from the woods along with the gray squirrels, which I could hear chattering in the branches overhead, scolding me—or perhaps it was the little red squirrel they were after, so devilish-looking with its pointed ear tufts, not much larger than a chipmunk (which I could also hear whistling their warning in the woods), who had braved both me and its larger cousins to purloin a peanut.

That's when I heard another sound, which at first I didn't identify, it had been so long:

"*Peek, peek, peek*"—then, softly—"*Tut, tut, tut.*"

I looked up. In the branches above me, having scattered wet snow on my head as they landed, was a flock of American robins, the orange-red of their breasts like flame in the frozen Aspen tree.

Their *peek-and-tut* call, I knew, signified a potential danger—not from a true predator such as a hawk, but from me, the hapless human being lurking below, looking rather disreputable in my bulky down jacket, stained with pine resin and torn from skirmishes with wild roses, and my ill-fitting flap-eared wool baseball cap, festooned with cockleburs and other stray sticky seeds.

Or perhaps, I thought, reconsidering as I started back to the house, my own warm world, still listening to the robins' insistent conversation, they were simply expressing their consternation. After all, the robins had flown in expecting spring and, instead, had found winter.

IT IS SOUNDS, MORE THAN SIGHTS THAT ALERT ME TO SPRING IN THE Catskills. The first time I hear the robin's song. The crunch of my boots in warm spring snow—footfalls are muffled in the dry powder of colder temperatures. The trickle of a stream finally escaped from ice, which will build to a rush during the rains of April. The return of thunder to the mountains, that first distant drum roll early in the season, as if announcing the awakening of the earth. It is the absence of sound that defines winter for me—not the total absence, as there is always the wind and how it can howl here, at night, especially desolate, like the call of the solitary coyote, which can carry for miles, or the *yip-yip-yip* of the pack, echoing off the mountainside, eerie and frightful.

Standing in the woods in winter, listening to the wind do its work, causing trees to creak and, occasionally, branches to crack, the report echoing like rifle fire in hunting season, or on the coldest days, when not even the crows caw or the chickadees converse, I even long for the *kreee-kreee* call of the red-tailed hawk, whose cry I dread once the spring migrations have begun, fearing for the songbirds. Finally, this day, my isolation is broken by a blue jay, a bold bird and clever too, known to mimic the screech of the red-tailed hawk as it flies into feeders, scattering the smaller birds and claiming the seed for itself, but in nesting season, keeping quiet so as not to attract the attention of other predators. Landing on a dead branch protruding through the fresh-fallen snow (from what I know to be the carcass of a black walnut tree blasted to the ground by a lightning strike last summer, whose warm days now seem so long ago), the blue jay eyes me curiously, its perky crest and handsome coat as blue as the winter sky on this rare sunny day, adding to its air of confidence.

Am I friend or foe? Am I the nice lady who trudges through the snow to fill the feeders on the edge of the western woods with sunflower seed, even scattering peanuts on the ground for him to steal from crows and squirrels? Or am I the madwoman with flailing arms who chases him and his family away from the yellow cedar tree where the robins nest every spring?

The jay cocks his head, staring at me intently, his small black eyes shifting into suspicion. He has made up his mind about me.

"*Jaay! Jaay! Jaay!*" he starts to admonish, "*Jaay! Jaay! Jaay!*"

His raucous call shakes the silent woods, snow falls from the branches above, as other blue jays fly in and take up the chorus: "*Jaay! Jaay! Jaay!*"

Then, having finished their accusations, the jays fly off as suddenly as they arrived, leaving me still in winter, alone once again in the silent woods, waiting for spring.

CHAPTER TWO

Miracles

(For Rachel Carson)

"THERE'S ALWAYS THE WISE GUY WHO'S UP BEFORE ALL THE others. Usually he starts singin' just before dawn, when it's still dark out. But if you squint your eyes, you can see that the sky's starting to lighten, turnin' to gray."

Farmer McCagg sucks on his pipe ("My wife says it's no good for me, but I'm still kicking"), pauses as if listening.

"One year it might be a robin, another year, a cardinal or one of those little phoebe flycatchers. That phoebe, one summer, woke me up every morning—'*pheebee-pheebee*,' over and over. What a damn racket that fella can make—woke up the cows, too, would get 'em all mooin'. I didn't even have to set my alarm. That's nesting season for ya, everyone calling for a mate, building nests, feeding their babies . . . The busiest and best time, May is, June, too. So much noise—in the woods, in the meadows—so much noise everywhere, after so much quiet."

He pauses, pulls on his pipe again. "Here in the mountains you don't need a calendar to tell you it's spring. I know when I hear the song of the first summer bird, come home to nest.

"No matter how old I get, hearing that always gives me a kick. There's no better medicine after a long winter, believe me, than birdsong. And I should know about medicine, I sure do take enough of the stuff . . ."

THERE'S ALWAYS A DAY, SOMETIME IN FEBRUARY, WHEN YOU VENTURE outside and, if you're fortunate enough to have a sunny day, you can see it. It's almost imperceptible, that first time, at least it would be to a visitor, perhaps. But when you've lived through a harsh Catskills winter—when you have felt the dying of the light as a deep personal loss and have known a cold so bitter, so penetrating that even for a few minutes, you risk frostbite if you venture out to fill the bird feeders, when you have known a wind that punishes, running up the mountain and hurling itself against the house, scouring the siding and peeling paint from the concrete foundation, leaving even the resolute metal flag pole with an irrevocable westward tilt—you can sense that change in the light. You can feel it even more than you can see it. It's as if the light itself is starting to grow—it's fuller, rounder, the sun on the ascent for nearly two months now, since December 21st, the winter solstice, the shortest day of the year. Just as there is that day in August when you step outside, the morning hot and bright, when the sun's light, so high overhead through summer, has started to slant, its rays angled, more diffuse. And if the wind is blowing (though it's a warm wind), underneath it is a coolness, subtle, quick, as slight as a breath. That, for me, marks the first day of autumn, no calendar needed.

In the fall, among the last birds to migrate are the robins, just as they are among the first to arrive in spring. That morning of the Big Snow, it was the sound of those robins overhead in the branches of the Quaking Aspen tree that truly alerted me to spring, assured me it would come, and forced me to look up when I spend most of the winter looking down, putting one foot in front of the other, marking the passing of days as slowly as I make that march out to the feeders in the deep snow.

MORE THAN ANY OTHER BIRDS, THE AMERICAN ROBIN AND EASTERN bluebird are icons of spring and, as such, have captured our imagination. Two of the earliest returning migrants—preceded in the Catskills only by an enterprising song sparrow and the tough, resourceful red-winged blackbird—both are songbirds whose voices, however different, are distinct for their melodious, sweet, and caressing songs. Their singing, as they proclaim their territory and court a mate, alerts us to spring, the season of desire for all species. But the first sight of their rich, warm colors against the chill bleakness of early spring—the bright orange-red of the robin's breast as it bobs along searching for earthworms, the bluebird's sky-blue back and tail feathers, its breast of red clay and belly of cloud white, a promise of summer to come—sparks within us an excitement as young as the season is new.

When I was in seventh grade, a classmate named Jonathan Meltzer (that I remember his name is indicative of how momentous this event was), rushed out of his seat to a window, where, pointing heroically, he proclaimed: "Look, a robin!" That revelation resulted in the first essay I ever wrote, which our teacher Miss Vogel (whose name, I would learn much later, meant "bird" in German), as straight and skinny as a stork, assigned us to write in class. Mine began, "Today Jonathan saw a robin. That means," I concluded—it was a two-sentence essay printed in block capitals on a short sheet of blue-lined paper—"that spring is here."

OF THE TWO, THE BLUEBIRD, A FAVORITE SUBJECT OF SONG (PERHAPS most famously, "Somewhere over the rainbow bluebirds fly . . ."), and often reduced, however affectionately, to a Disney cartoon caricature, has always had more caché than the robin, our most common (faint praise, that word) and largest American thrush. The bluebird is not just smaller in stature than the robin; it is slender as opposed to chunky, slight as opposed to sturdy, and more elegant than stalwart in appearance. Robins, which tend to flock together, also give, as a result, the impression of gregariousness, whereas the bluebird can appear shy, favoring only the company of its mate and offspring. Even its song, so lyrical and soft, is subdued in comparison. The bluebird is also rarer, in that it favors the countryside, seeks open spaces with woods offering tree hollows and fields with fence post holes for nest building, while the robin belongs to all environs, from city gardens to suburban parks, from country meadows to mountain valleys, making its nests in bushes and tall shrubs as well as trees and manmade structures too, the eaves of roofs and even, one spring, atop a neighbor's porch light, dismaying my friends who had put out "such lovely boxes for them, well sheltered, with easy access to come and go." It was a quixotic choice that forced them to leave their front porch in the dark through two broods, until all the nestlings finally fledged.

And then came the crisis in the bluebird's very existence—threatened with extinction by the loss of habitat especially, the felling of dead trees and removal of branches that, in the past, in rural settings, would provide them with nest hollows. Perhaps another reason that we prize the bluebird so, treasure even a glimpse and see it as a benediction, is that their recovery in numbers (their population fell by almost 90 percent at the height of their decline) has rebounded since the 1960s to the point where they are no longer threatened as a species, thanks to that magical, thoroughly practical invention known as "bluebird boxes," which have allowed bluebirds to nest and now maintain their numbers. Thomas E. Musselman, generally regarded as the founder of the bluebird conservation movement, began making nesting boxes as early as 1926 in Adams County, Illinois, erecting them in "bluebird trails" (a series of nesting boxes, often in a line or circle), also his concept, in response to the increasing threat to bluebirds posed by marauding starlings. This inspired succeeding generations of dedicated individuals, from boy scouts to backyard birders and, most notably, a poultry farmer named Ralph Bell, who in 1964 put up a bluebird trail of some two hundred boxes on utility poles along country roads where he delivered eggs in southwestern Pennsylvania with as many as 800 bluebirds fledging each year. In 1999—in a stunning, rare and welcome reversal—the Eastern Bluebird was removed as a species of special concern from New York State's Endangered, Threatened and Special

Concern list. Bluebirds are a success story, one that we can largely attribute to human intervention.

But as much as much as I have a passion for bluebirds, I want to
argue here for the robin: not just as a popular icon of art, song, and poetry
or a "harbinger of spring," but as representative, for me, of nature's perfection, as a shimmering strand of the web of life.

First, I speak for what robins have done for us. (What critters
do for humans always seems to increase their value.) In the 1950s and
'60s, poisoning by the spraying of DDT to combat Dutch elm disease
was instrumental in generating concern over a potential "silent
spring"—DDT-coated elm leaves eventually processed by earthworms,
which, along with contaminated insects, were then devoured by robins,
led to death or reproductive failure in the birds. Birds are at the top of
the food chain, as we are. Raptors especially—the peregrine falcon, osprey, and even our national bird, the bald eagle—were also affected by
DDT, most notably in the thinning of eggshells, which broke during
incubation, causing the death of offspring. DDT had filtered into
water, where it was absorbed by fish—the osprey and bald eagle's primary food—and seeped into the earth. Rachel Carson, in her seminal
book that effectively sparked the modern American environmental
movement (and contributed to the founding of the Environmental Protection Agency in 1970), *Silent Spring,* published in 1962, took note of
these correlations and was vilified by the chemical industry and its al-
lies in Congress. (The title of Carson's book, *Silent Spring,* is a haunting
reference to a world without birdsong—a "silent spring" empty of
birdsong, that joyful noise of renewal and life.) A rebuttal to her book,
"A Scientist Looks at *Silent Spring,*" published as a pamphlet by the
American Chemical Society, charged that Carson's view would mean
"the end of all human progress. . . . It means diseases, epidemics, starvation, misery and suffering."

Eventually, though, Carson's claims were validated and DDT was
banned in the US on June 14, 1972, which is considered the first major

success of the environmental movement. Still, species that were endangered—peregrines, osprey, bald eagles, as well as brown pelicans—rebounded significantly within twenty-five years, and, at least from our perspective, Americans can eat, drink and breathe more healthfully, and our children can live in a cleaner environment as a result of the ban.

So the robin, which is so often found near humans, a resident of both cityscape and countryside, has also been a boon to us as well. The proverbial canary in the coal mine that warned miners of the presence of noxious, life-threatening gases, giving its own life so they could escape danger, is a role that birds play for us today, as the robin warned us in its sacrifice of the dangers of DDT. But as crucial as the discovery of the harmful impacts of such poisons on the environment was, just as important has been our resulting awareness of the connectedness of life—to wit, that what an earthworm eats can impact even our own vaunted human drama. Nature is in such perfect balance, and the connections that comprise it are astonishingly simple, yet utterly sublime, rational yet magical, too, subtle and obvious at the same time. They are, in a word, wonderful, which is how I feel when I contemplate the spring migration of the American robin—full of wonder.

The robins migrate along what is known as the "37-degree isotherm"—that is, an imaginary line along which the temperature, as the line heads north, averages 37 degrees, day and night. What this correlates to is that earthworms, which migrate vertically to the surface, begin that journey at 36 degrees. Thus, robins, which usually travel north following a warm front and arrive with spring rains, are assured a welcome protein feast upon their arrival. (Worms, of course, will be a primary food—for their protein value—served to fast-growing nestlings.)

But what of the robins who arrived that March morning of the Big Snow? Robins are large enough birds with, at that time of the year, sufficient fat reserves to survive, if they need to, for several days. They are also fruit eaters, especially in the fall (when worms travel down into the earth to escape below the frost line) and early spring (if the worms have not yet been lured back to the surface by warming temperatures). They know to head straight for fruit trees, cherry and apple here, as well as mulberry, elderberry, and blueberry bushes—staples in the Catskills, not just for birds, but for black bear, who gorge on fruits and berries in fall especially, fattening up for hibernation and making "blueberrying" on the mountain an activity that requires caution, stealth, and sometimes speed. In the case of my flock that day, I didn't see them again until the snow started to melt, leaving the ground soft and exposed, ready for worms. I knew the robins had probably turned around and headed south to wait out the elements—robins are "facultative migrants"; that is, they can choose to migrate or not, depending on circumstances, as opposed to those that must migrate to survive (hummingbirds, for example). As a result, they migrate only as far south as they need to, when forced by bad weather or shortages of food. (Even in northern climes, if robins, hardy birds that they are, can find enough to eat they will stay through the winter, though I have not yet seen them in winter in the Catskills.) The Big Snow had not only buried the ground in almost two feet of snow but sleet at the end of the storm had glazed whatever fruits and berries were left after such a long, harsh winter.

Migrating in flocks has its method, too. First, it allows for greater safety from predators. It's easier for a hawk to pluck a stray flying on its own than to penetrate a tight, fast-moving flock. (European starlings particularly have mastered this stratagem, flying as one to deter predators, shape-shifting dramatically as they swoop and dive together.) And in a flock, more experienced birds will help less tested ones, especially adolescents migrating for the first time, by showing them food sources as they no doubt did one fall when wave after wave of robins flew into my meadow, staying as late as November to feast on a boon of apples occasioned by a very wet summer. When my flock did return, I noted, as before, that they were all males—their darker, almost black head, as well as bright reddish-orange breast, set them off from the females, whose head is paler, brownish-gray, and whose breast is orange, but duller. As is the case with many migratory birds, the males precede the females to establish territories, which the females can then evaluate, along with the males' singing voice, choosing a mate for his property and sex appeal (the everlasting appeal of the crooner!), just as humans often do. (The male's plumage can also be a persuader—good grooming seems to count in many species.) Female robins follow within two weeks of their prospective mates when the weather is warmer, when rain instead of snow ensures that there will be mud for nest making, a vital component of the fashioning of a robin's nest. Both the male and female robin attend to the nest, and the male robin, as the male bluebird, will care for the fledged first brood, tirelessly feeding his demanding offspring as the female incubates their second clutch.

IT IS NEARLY MID-JUNE—JUNE, THE MONTH OF BABIES, AS I CALL IT. Just this morning, in the meadow, I glimpsed a doe nursing her newborn fawn who stood on wobbly legs as her mother devotedly licked her bottom to stimulate urination and defecation, which is necessary for the baby's normal bladder and bowel function. The wild turkey hens, hiding in the meadow's tall grass, their long necks emerging like periscopes, are also hiding their poults, whose soft *peep-peep-peeping*—and their mother's clucks, commanding them to follow—give them away. Flightless for the first two weeks of life, the chicks, who bumble along as they try to keep up with their majestic mother, roost beneath her wings at night, in effect, hiding behind her feathery skirts from their world's many dangers. The Eastern cottontail kits have started to appear too, already on their own after three or four weeks—so small and seemingly guileless, it's amazing to me that any survive. At dusk, several emerge from my wildflower garden, where they hide during the day, to nibble on fresh-cut grass, tolerating my presence too well. Their parents' courtship is a spirited minuet of choreographed steps known as "cavorting"—the buck rabbit chases his doe until she turns and boxes at him with her front paws, then they take turns jumping straight up into the air. Dance, it seems, is necessary for dating in more than just our species.

The other day, one of my cats—who don't go outside, as much as to save them from coyotes as to spare the songbirds their coarse attentions—was sitting where she shouldn't have been, on the dining room table, her body quivering, chattering her low hunting call. What Hannah was pointing at, as tenacious as a bird dog, was a newly fledged robin, perched precariously on the wooden rail of my front deck, plump and fuzzy, paler than its parents, with the adolescent's signature dark spots on its breast. The youngster wore the same stunned expression I have seen on other newly fledged birds—a pigeon squab raised in a tight corner between a wall and an air conditioner on the windowsill of a New York City apartment, its only view until it took its first flight to the fire escape, a brick wall; a bluebird newly released from its dungeon, the dark, humid nesting box that had been its home until its parents coaxed it out, finding, that time too, the target of my deck rail, where it sat for some time, looking around, perhaps contemplating its own courage. The young robin had been raised in a nest beneath my front deck built on a ledge provided by a support beam. Its view of the world had been cast in shadow, with only the concrete foundation of the house to gaze on, and through a mud-splattered basement window just opposite, an occasional human face—my own—as I checked every morning to make sure the robins had survived another night, the nest being only three feet off the ground, and I have seen a red fox patrolling these grounds at dusk. I had left the porch light on through the night, hoping to deter the fox as well as the raccoon twins I spied through my bedroom window at midnight, looking like robbers with their black masks, which I frightened away with a shout. Now, this was my reward—the sight of a new robin wondering at a world that was new to it as well, a *wide* open world of light and color, and, a warning from the cat, a world of danger, too.

All four robin nestlings fledged, as I was delighted to discover later that afternoon when I peered through the window at their empty nest.

But that basement window would only stay open briefly, as, after a few days, when both parents cared for the fledglings, the mother robin returned to the messy nest she had fashioned of mud, grass and twigs, adorned with a long stick that hung down from the outer layer like a rustic bell cord, to start her second clutch of the season. Robin père was now charged with feeding his babies as well as protecting his new brood-to-be. Even with his divided duties, he was a stalwart defender, sitting in the spindly Japanese maple tree that I misguidedly planted on this rocky mountain (a gift from equally misguided friends) and have kept alive through harsh winters and false springs, expressing his consternation— *"peek, peek, peek–tut, tut, tut"*—when I walked out on the front deck, even though I assiduously avoided the occupied end, not wanting my footfalls to frighten mother robin. I didn't linger, didn't want him to be upset—and I didn't want to be dive-bombed either. I am always amazed at the bravery of birds, of mothers and fathers, who defend their nests and their young so fearlessly. This poppa robin had taken off after much bigger crows and more aggressive starlings—prudent that, as those species will take robin eggs, even young.

But sometimes you just have to laugh. Three chipmunks, all called "Scout" (since I can't tell them apart) frequent my back deck, feasting on sunflower seeds that I leave out for them. One of them, I noticed, had scampered around the house, disappearing under the front deck, and I suppose I shouldn't have been surprised when I peeked through the closed basement window, wanting to discreetly check on robin mère, and found myself nose-to-nose with the chipmunk. Both of us were so startled, we both jumped back. But it was later that day, when I saw an even more comical sight. The chipmunk had wandered again into the robins' territory—and this time, robin père finally noticed. I saw the robin literally riding the fleeing chipmunk's back, beating it with his wings. It looked like a cartoon or a scene out of some "incredible journey" movie. But for the robin, this was serious business. As "cute" as chipmunks are, as easily as they charm us by taking sunflower seeds from human hands, they will also take bird's eggs. They are small and clever enough to hide near the nest until the mother leaves it. This time, though, the message was delivered. From then on, "Scout" stayed away.

What makes that instinct —so powerful that a female robin will attack a redtailed hawk that has downed her mate, then grieve over his body until it is removed (as I had to do)—so powerful that a male robin will defend his family against such formidable adversaries, including me, disregarding the risk to himself, so powerful that both parents will work tirelessly, from dawn to dusk—in rain, in wind, in heat and cold— in perfect concert, to feed their young? We call it "instinct," an innate behavior that needs no learning, geared to species survival, and treat it as less evolved than our own refined emotions—in particular, the bond we call "love." But our own human bond, even and perhaps especially in families, can be absent, erratic, and all too often, destructive. While "instinct" in these critters, this most "common" of birds the American robin, I have seen as loyalty, as devotion, as courage—which is what I would call love.

How did it evolve, the robin's reliance on the earthworm? That delicacy of the 37-degree isotherm, predicated on worms rising to the surface at 36 degrees? The fact that female robins hold back their migration north until mud for nest building is assured? These simple facts of the robin's spring journey, when taken together, become sublime. They point to the complexity of life, and as part of that, its interconnectedness —in the robin's case, not just to other animals, earthworms, of course, lowly only in their proximity to the ground, but to light, to weather, even to mud. That interconnectedness is what we refer to as the "web of life"—cliché or not, it describes it best, the mystery and, for me, life's many miracles.

"*Bzzzt!*" THE FIRST TIME I HEARD THAT BUZZING, ALMOST-NASAL sound, I thought it was my back doorbell. *"Bzzzt!"* Then I raced downstairs to the basement of my new Catskills home, to see if some alarm indicating one of several possible catastrophes—electric? water? propane?—had been triggered. *"Bzzzt!"*

The sound was so odd—"It couldn't be a bird," I told myself. "What kind of bird sounds like a buzzer?"—I finally figured out, was outside, coming from my nearby woods.

"Bzzzt!" It repeated every few seconds, then after several minutes, no more—the "buzzer" abruptly stopped working.

The woods were not silent, though. As I stepped out on my back deck that mild evening in April, a mist rising over the meadow from the patches of snow still scattered on the ground, I was met by another announcement of spring on the mountain—the sweet jingling song of spring peepers, the tiny, cryptic tree frogs that are rarely seen but whose voices, trilling together, full of silvery desire, the males pleading for a mate, enveloped me as I stood in the settling dusk. Then it began again—*"Bzzzt!"*—a rather rude punctuation to the peepers' ardent, crystalline calling—*"Bzzzt!"*—and continued at the same set intervals as before, until darkness finally stilled this strange new sound.

Becasse IS WHAT THE WOODCOCK IS CALLED IN LOUISIANA, THE SOUTH-ernmost part of its American range, which spans the eastern half of the U.S. to the Great Plains. *Becasse,* a Cajun term of endearment for this popular small game bird, the quixotic and curious woodcock, is from the French *bécassine,* which means "snipe." Just saying the word "snipe" can elicit laughter at what is deemed by many to be an imaginary crea-

ture, the illusory object of that practical joke known as the "snipe hunt."
But there is a snipe that *is* real, any one of twenty-five species belonging
to the shorebird family *Scolopacidae*, which also includes sandpipers and
their country cousin, the woodcock. And here is the first of several pe-
culiarities of the woodcock—it is a shorebird that has adapted to live in
shady forest uplands, where it probes the moist soil with its strong bill
for earthworms, its insistent digging reminding me, when I spied one at
work in a meadow this spring, of least sandpipers, the "peeps" that skit-
ter along the ocean's edge, busily excavating the hard wet sand left exposed by retreating waves for tiny crustaceans. The woodcock's
shorebird affinities—its round, plump body and brownish-gray coloration, as well as the way it hunts for prey, stabbing at the ground with
the long, slender bill that is its most distinguishing feature—would seem
to be reflected in the fact that of the many nicknames evoking its habits
and habitats, and proving the bird's oddball appeal, several contain
"snipe": blind snipe, brush snipe, cane snipe, dropping snipe, forest snipe,
owl snipe, and wood snipe.

But I also suspect that the woodcock's snipe alter ego is, in part,
the elusive "snipe" of legend, that bane of hunters of every age and ex-
pertise, including me, when I searched for "snipe holes" with other kids
on summer hikes led by Major "Barney" Bernard, the father of my child-

hood best friend Peggy, who'd spin yarns of courage and derring-do, invariably set in Occupied France, reflecting his own wartime exploits, which managed to include as characters every one of us, boys and girls alike.

"And then Bobby pushed open the door to the hideaway, his sidearm drawn. He squinted his eyes, trying to see in the dark. That's when he heard it—the low hum of voices . . . Was it friend, he wondered, so far behind enemy lines, or foe?"

"Bet it's Nazis," Russell interrupted, "you're done for, Bobby" (though no one ever died in these epics—not even Nazis), as the rest of us chorused, "Then what, Major Barney? What happened next?"

"To be continued," he'd say, rising up from our "campsite"— whatever rocks, logs or stumps we had happened on, on our hike—"at the next snipe hole."

Then we were off, searching high and low for the snipe's secret lair. "Don't run," Major Barney would caution, "you'll fall into one." And any time the cry of "snipe hole!" went up, he'd examine the find and either dismiss it—"No self-respecting snipe would live *there*"—or nod approvingly and reward the child who'd made the discovery with the coveted "Well done, Sergeant." Of course, Major Barney's capricious choice of snipe hole—no two ever seemed exactly alike—only came after he'd had time to think up the next installment of our story.

"So Bobby started down the stairs—they creaked and he stopped short. Then, in the dim light, he saw them . . ."

"Who, Major Barney?" we shouted, "Who?"

"Those two fearless fighters for freedom, heroes of the Resistance"—he paused, looked at each of us, one by one, finally his gaze settled on—"Leslie and Peggy, sitting at a table, playing poker and smoking cigars. 'Hiya Bobby,' they called out, 'pull up a chair.'" (How wonderful for me, a girl growing up in the '50s, to have the possibility of being a hero, let alone playing poker and smoking cigars!)

Later that day, I mustered up my "hero's" courage to ask Major

Barney what my mother would have described as the "sixty-four dollar question," as we sat at the counter of Foster's Drugstore, where he always took the "troops" for ice cream sodas at the conclusion of every hike: "Is there really such a thing as a snipe?"

Major Barney laughed and said, "We wouldn't be hunting for snipe holes if there was no such thing as a snipe, would we?" For a moment, his lawyer's logic stumped us.

Then Bobby, the same "soldier" who had braved the "hideaway" of today's story, boldly asked: "Then why aren't we looking for snipes?"

Major Barney, always a jolly man, turned quite serious. When he answered, we leaned in to listen as we did when he spun his tales of war. "Only people who are very lucky ever get to see the snipe."

"Why?" Lynde blurted out.

"Because," he said slowly, "it can make itself invisible. And even if you are fortunate enough to see it in the woods, where it lives, you can never catch it."

"Bet we could catch it," Bert insisted.

"Can't catch it if you can't see it, and you'll never catch it because it sees you first." This was met by silence. "The snipe is the only bird that has eyes in the back of its head," Major Barney announced. "Its eyes are so big, as a matter of fact, that it can see in every direction, front as well as back, sideways, too. You can't sneak up on a snipe"—he shook his head—"You'll never get the drop on it."

Major Barney's description of the snipe as the bird that sees everything but is never seen, which he added to on subsequent hikes—no doubt in an attempt to deter us from rushing deeper into the woods to find it—"It has a bill as sharp as an officer's saber" and "If you step too close to it, it will explode like a grenade"—would be lost in my memory until I happened on what is for me the most peculiar snipe nickname for the woodcock—*dropping snipe*.

"'Dropping snipe,' as in, can't get the drop on it," I'd laughed, and it was then that Major Barney's words, including that phrase of his, often invoked when discussing military strategy in our stories—"Gotta get the drop on the enemy before the enemy gets the drop on you"—came back to me. That long-ago snipe of the snipe hunt and snipe holes, had sparked my lifelong interest in birds, my fascination, really, with them, my sense of them as almost magical creatures, with their own private lives and secret languages, and though I no longer looked for the snipe of my childhood hikes, I still sometimes startled friends while walking in the streets of New York City by suddenly calling out "snipe hole" at the sight of a storm drain.

Recollecting Major Barney's description of this baffling bird, the Holy Grail of hunters and seekers after subterranean holes, I wondered if the snipe of lore was, in fact, real—not one of the twenty-five shorebird snipes in good standing, but their upland relative of many names, *Scolopax minor*, the American woodcock.

Big eyes, big-eyed John, and *owl snipe* (reprised). The woodcock has large, dark eyes set high and far back on the sides of its head so that it can see all around, including behind itself, even with its head down, in low light, as it digs for food. Its "eyes in the back of its head" is a unique adaptation (true to Major Barney's assessment), which also allows the woodcock to see what is above it—especially useful for a bird that is subject to the unwelcome attentions of hawks by day and owls by night—as it doodles along the forest floor (*timberdoodle*, the woodcock's best-known nickname).

The woodcock's coloration is what renders it "invisible," its bluff hues, soft yellows and grays, and broken black and brown "dead leaf pattern" camouflaging it perfectly in the litterfall of leaves, twigs, and bark. The woodcock stays still and hidden if danger approaches, but if you get too close, it will indeed "explode like a grenade," erupting into flight at your feet, the quick beat of its short, rounded wings—designed to navigate the thick forest understory—reminiscent of the skillful aerial moves of that other adept flyer, the bat (hence, these two nicknames, *mud bat* and *swamp bat*).

The Seneca Indians, bemused by the woodcock's peculiar appearance, offered this legend as explanation: the woodcock, they believed, was made by the Creator out of the leftover parts of other birds. Indeed, its head looks too large for its body (on which it seems to sit without benefit of a neck), and its eyes and bill look too large for its head. Standing around five inches tall, and measuring ten to twelve inches long, the woodcock is about the size of a robin, albeit a chubby one. Considering that the robin's bill, which is long and slim relative to other songbirds' (and which, like the woodcock, it uses to probe for worms), is less than half the length of the woodcock's, which averages two-and-a-half inches, the bill of *bec noir* (another Cajun name, meaning "black beak") is prodigious in comparison.

But as ungainly as that large sharp "saber" set in that small round head may seem, the woodcock's bill is perfect. Its shape and size make it an ideal tool for probing the ground, especially the soft, damp mulch of shaded forest spaces, to ferret out food. As the woodcock hunts, sensitive nerve endings in the bill's tip alert the bird to any worms or insects—elusive ants or quick-stepping spiders—as well as those crunchy centipedes and tasty Crustacea, snuggled in the soil. When it locates its prey—especially the staple of its diet, nutrient-rich earthworms, smooth and sleek, which are burrowed below the surface—the flexible tip of the "prehensile" upper bill, or mandible, can be opened by the bird while its bill is still submerged in the earth, searching for supper. That upper mandible, as well as its long tongue, which then emerges, has a rough surface that enables the woodcock to clasp these slippery, stubborn critters and pull them out from their hiding places underground, to enjoy its own al fresco buffet. *Bog borer* and *bog sucker* are nicknames that pay tribute to the woodcock's skillful foraging maneuvers, and particularly, to its signature feature, its disproportionately large and seemingly clownish bill, which is, in actuality, an adaptive masterstroke, a marvel of form fitted to function.

And finally my favorite, *hokumpoke* . . .

The first time I saw a woodcock walking in my upper meadow, no doubt searching for worms in the wet spring earth, its rhythmic moves reminded me of dance steps. Which made me wonder if its *hokumpoke* nickname didn't derive from that '50s dance craze, the hokey-pokey. But when I did my due diligence, I discovered that the hokey-pokey ("You put your right foot in, you put your right food out / You put your right foot in, and you shake it all about . . .") was *not* the dance the woodcock was doing. It seemed to me to take two steps forward, then one back, rocking its body back and forth as it walked, its head held still. That dance reminded me more of a cha-cha-cha (with a hint of rumba in the woodcock's hips).

This so-called dance that the bird does as it ambles along is so charming and disarming and just plain funny that the woodcock has become something of a YouTube celebrity. Avian photographer Alan Stankevitz has posted online a video of a woodcock cha-cha-cha-ing its way across a country road (and taking its time doing so), to the accompaniment of "Walk Like an Egyptian," the 1986 pop hit by The Bangles. The film (which may be shorter than its credits) is hilarious, guaranteed to make you laugh and get you dancing, too.

One revealing aspect of the woodcock's walk captured in Stankevitz's close-up is how it slaps the ground with its front foot as it steps. This motion may well be intended to roust worms lurking in the soil beneath the bird's feet, scattering them, their subsequent movement making them easier to detect. And its seeming back-and-forth walk—that odd bobbing gait—which also aids in flushing prey, is the result of yet another adaptation—how the woodcock's short, pinkish legs are configured, far back on its body.

What an amazing little bird this is, that it can delight us so. The naturalist may marvel at the woodcock's many incongruities, how they fit together like the pieces of a mismatched puzzle to create a perfect picture—a critter that is so well suited to its environment. But the rest of us can always just enjoy the woodcock's cha-cha-cha, stepping and rocking as it hokumpokes along, walking like an Egyptian.

"*Bzzzt!*" THE NEXT EVENING WHEN I AGAIN HEARD THAT STRANGE buzzing sound, I decided to follow it into the woods. *"Bzzzt!"* It was still light, but heading toward dusk. I have the natural, perhaps necessary, human reserve about entering the forest at night. But armed with a flashlight and a walking stick to steady myself on the uneven, rocky terrain, I took the Cut—a natural opening into the woods just a hundred feet or so from my house—and found myself at the small, spring-fed brook where I paused, trying to fathom the direction of the *"Bzzzt!-Bzzzt!"* as it seemed to ricochet off of the trees and the mountain itself.

As I walked, tenuously, mindful of the darkening sky, I took the precaution of marking my way with pieces of loose bluestone—always plentiful, as the mountain is made of it. I was hesitant because I didn't yet know these woods, having moved here recently, and the darkness could be not just intimidating but disorienting, too. As I slogged through the mud of the April snow melt, which fought to capture my Wellington boots, causing that loose sucking sound amplified in the quiet of the forest, the buzzing suddenly ceased. But I kept walking in the same direction, following the stream, until I arrived at what appeared to be a clearing, dotted with shrubs and saplings, which had perhaps once been a cultivated field, a remnant of the farmer who had owned this land long ago. Then, I heard it again—*"Bzzzt"*—and several yards away, at the clearing's brushy edge, I saw the source of this curious, insistent sound— a small, plump, brownish bird with a prominent bill, which I knew from the delicate drawing in my Peterson guide to be an American woodcock.

Suddenly, the woodcock took flight, launching itself straight up, startling me, as it spiraled higher and higher into the twilight sky . . . As it flew, I could hear a melodious twittering sound, which, I would learn, was the whistle made by wind passing through each wing's outer three primary feathers. As the woodcock continued his climb—easily more than two hundred feet or so—I struggled to see him in the dim light. Then he seemed to level off and started to fly in widening circles, directly above where I was standing. Abruptly the bird fell, so precipitously that I gasped, fearing that he had been shot. But the woodcock was in artful control, zigging and zagging as it descended, all the while chirping so sweetly. Then he landed silently, practically on the same spot from where he had taken off. Almost immediately, he began his buzzing noise again, but this time, I noticed, I wasn't his only audience. Another woodcock had emerged from the undergrowth, barely visible at the brushy border, a female, I reasoned, as I had fathomed that this must be a courtship display. It was an intimate moment in the life of these birds and I withdrew, having to turn my flashlight on to find my way back through the shadowy woods.

The legendary sky dance, the male woodcock's mating display, accounts for the woodcock's most storied persona: *sky dancer.* The celebrated ecologist Aldo Leopold translated my *"Bzzzt!"* interpretation of the woodcock's call into the more familiar *"peent,"* coining the noun and verb *"peenting."* He has also written this description in his classic of conservation *A Sand County Almanac,* which made the sky dance familiar and famous:

> Up and up he goes, the spirals steeper and smaller, the twittering louder and louder, until the performance is only a speck in the sky. Then, without warning, he tumbles like a crippled plane, giving voice in a soft liquid warble that a March bluebird might envy. At a few feet from the ground he levels off and returns to his peenting ground, usually the exact spot where the performance began, and there resumes his peenting.

There is something so moving about this sky dance, even more so that it is performed by such an unexpected hero—the squat, awkward-seeming (at least on land), low-to-the-ground, funny-looking, even drab little woodcock. To see him fly so high, sky dancing in the dusk, executing his extraordinary twists and turns, is to be inspired and reminded that within each of us—as unassuming as we may seem, set in our daily routines, so that for the most part, we go unnoticed—is the potential for brilliance, for some moment of transcendence, when we rise above the everyday and become transfigured by our own specialness, the uniqueness that is within us all.

THE AMERICAN WOODCOCK IS A MIRACLE OF ADAPTATION. EVERY awkwardness, every eccentricity of appearance—how the woodcock is made—has a purpose. I can understand these adaptations, their necessity, the sublime inner logic of each—how the woodcock's cryptic coloration hides it in its habitat; how its supersized bill is designed to detect and seize prey; how its backset eyes and legs, which make the woodcock such an easy caricature, are keyed to its survival. The "how" that does remain a mystery, is how these changes occurred, over what time period—and how this seaside resident came to be a woodland critter, how a shorebird snipe transformed itself into the *timberdoodle*.

But these secrets of the woodcock's past, as intriguing as they are, do not compel me as much as this mystery of its present: How to explain the male woodcock's metamorphosis every spring into *sky dancer,* that this landlubber—built to fly in low, short bursts, through forest underbrush and branches—would turn into an acrobatic aerial artist to woo his lady loves. Nature, in its capacity for irony, and what can sometimes seem to us, humor, seems to have outdone itself with the woodcock's spectacular display. But then, what better way for the woodcock to distinguish himself from the rest—quite literally, to rise above the competition—than to show off how high he can fly, how compelling his spirals, how death-defying his sudden plunge, how smoothly he lands, "sticking" the same spot from which he took off? Or is it his courtship song that the female finds so alluring—the *peenting* that resounds through the woods at dawn and dusk, or is it his other "calls"—the twittering of his wing whistle as he flies high into the sky or then, as he makes his dramatic descent, his chirping accompaniment, so sweet and insistent—that steals the hen's heart? The truth is that we can't really know what the female woodcock looks for in a mate, how she chooses one earnest, hard-working male over another—woodcock chemistry, and attraction, could be individual, as it is with humans, as well as her own closely kept secret.

We do know that the chunky, ambling woodcock metamorphoses every spring into the swift, graceful *sky dancer* to appeal to his prospective mates. But is everything really so mechanistic, is everything just a rote, instinctive pattern of behavior meant to ensure procreation, the survival of species? Then why does the *sky dancer* "dance" even after he has finished mating and the hens have successfully brooded their young—why do the males continue to "display," flying high into the sky, utterly exuberant, throughout the summer? These may be adolescents, one theory goes, "practicing" for next season, and perhaps some are.

But I believe that the woodcock also dances for another reason—the same reason that every spring, there is that one robin that will start his *cheerily-cheer-up-cheer-up-cheerily* song before dawn etches color into the gray morning sky, that will sing into deep dusk, in defiance of the onrush of night, reluctant, even for a few hours, to embrace silence—I believe that the woodcock also dances for pure joy.

TREES, SO SILENT, IMMOBILE, ARE TAKEN FOR GRANTED AS LIVING things. Are they sentient?—perhaps not in the way that we know it. But they are companionable, I feel so strongly, in my walks in the woods, leaning against the trunk of a sugar maple, shielded from the sun by its leaves in summer, which turn crunchy underfoot in fall as I hike, such a hardy sound compared to the lilt of a single leaf as it glides gently to earth. Everything about a tree is alive. In spring, on the mountain, the blush of red on branches—at first, barely perceptible, more hoped for than real—alerts us to sap starting to flow, the life blood of these winter-weary trees beginning to circulate once again, as days warm and nights, still chilly, are no longer frigid.

A yellow haze soon appears—a texture as much as a color—the fuzziness of buds beginning to emerge on bare branches. As the buds unfurl into leaves, the colors shift into a subtle array of greens, each species of tree slightly different from the next, but all are tender, delicate, dreamy even, reminding me, as I look out on this scene of the mountain in spring, of an impressionist painting, so fresh and new, yet timeless too. Even "dead" trees are alive, providing homes for birds such as woodpeckers and bluebirds, which nest in their hollows, and in winter, fallen boughs and toppled trunks may shelter hibernating bears—as I discovered when I happened on a sow and her two cubs asleep in the snow, in a den dug under a fallen black walnut tree, a sight that thrilled me, my heart pounding, as I slowly, as silently as possible, backed away into the woods (and, I confess, went straight home!). If you take the time to look very closely at the trunk of a winter tree in the Catskills—maple, aspen, oak, beech, birch, red cedar, white pine—you might be fortunate enough to find, hidden beneath a loose piece of bark, yet another miracle—a slumbering tree frog, that noisy harbinger of spring now silent, the spring peeper.

Of the wonders I have chronicled here, the spring peeper is perhaps the most intriguing, secret and even endearing. It is, for starters, certainly the smallest—roughly the size of one's thumb (and when photographed next to a paperclip, it is barely bigger)—its length varies from less than an inch to an inch and a half, and its "weight" is scarcely that, from 0.11 to 0.18 ounces. Obviously, as small and as camouflaged as the peeper is—its tan, brown, gray or olive drab coloration mimics tree bark, its favorite hiding place—in its woodland habitat, especially in summer, the peeper is hard to spot. What is extraordinary is that in April, here in the Catskills, but elsewhere in the East around the time of the vernal equinox, after the ice has started to melt, when snow has shifted to rain, filling wetlands, waking the marshes and ponds which this little amphibian requires to breed—when that first trilling high-pitched whistle is heard, at dusk or deeper into the night, as the weather warms, a chorus of these frogs begins, rising in velocity and cadence, the truest, fullest, most utterly triumphant ode to joy I have ever known—one can still never find one.

At night, as that is when they mostly sing, staying under cover of low shrubs and tall grasses, which they cling to with large toe pads that serve as suction cups for climbing and hiding from their many predators (including even certain beetles, these frogs are so tiny), if you venture out, finding yourself in a wet spring meadow, surrounded by their sound, which, when they sing together, has been likened to sleigh bells, even with a flashlight seemingly pointed right at them, they will elude you. Sometimes they will simply fall silent, sensing your presence. But from where they hide—including cracks in the ground and under shrubbery—they can amplify and "throw" their voices, in effect becoming amphibian ventriloquists, impossible to detect ("fairy frogs," my niece calls them). Their call—about one per second—a rising, high-pitched peep performed by males at the edge of pools of shallow water—grows more intense the more bachelors there are competing for their females' favors. When the peeper female chooses her own crooner (again, the power of song in so many species, including and perhaps, especially, our own, to attract—from The Beatles to Justin Bieber), she will lay, at a time, approximately a thousand eggs per clutch in vegetation in ponds.

The birds of spring bring life to the dawn and dusk of our lengthening days. But to hear the peepers—when it's dark and still so chilly that you can see your own breath—start to sing, even just one peeper peeping, is transcendent. Perhaps because they are heard but so rarely seen—at least not by me, despite my walks in the woods searching for them, stepping carefully in leaf litter in summer, scrutinizing the bark of trees in winter— that the peepers can seem magical. They are so welcome, singing as they do, through May and even into June in the Catskills, until they have successfully discharged their duty to procreate, when the green frogs join in with their own summer song. The affection we have for the peepers can be shown by our nicknames for them—"pinkletoes," "tinkletoes," "pinkwinks"—which all try to capture their cheery, incessant calling. It is a joyful noise, perfect as a psalm, reverently produced by a sac in the male's throat, which expands and deflates like a child's balloon with each "peepy toad's" peeping. There is a Catskills saying that when you've heard the peepers singing for the third straight night, that's the charm—you know that it is finally spring. The peeper will peep even when temperatures dip into the thirties. But on those cold, early spring nights, when the temperature, resisting the calendar, dips below freezing, these chorus frogs will fall silent. In high winter, they hibernate, their hearts beating very slowly, barely breathing. But the tree frog, like all amphibians, is cold-blooded. Their bodies, in part, freeze—but usually only in part. In an extraordinary adaptation, the peepers produce glucose in their livers—glucose is a sugar that acts as a sort of antifreeze—which is pumped to the frog's vital organs such as the heart and lungs. Other parts of the peeper, its extremities, form ice crystals. In this particular frozen state, it hibernates under leaves, in the earth, and beneath loose tree bark. In early spring, temperatures finally rise above freezing, thawing the frogs, warming them enough to awaken from their slumber—and the temperate spring rains tell them it is time to sing again.

The triumph of this tiny tree frog, the spring peeper, its miracle of miracles, is that it survives—not just winter but so many predators—bugs, mammals, bats. But the peeper is a predator too, hunting insects, especially mosquitoes, in its nighttime work—a benefit, that, to humankind. A peeper chorus can be heard for over a mile away, depending on the number of suitors singing. There is no sound I love more in spring, no sound that affirms life, new life, as this does, in defiance of winter's long night, of the cold—even the freezing cold—that may daunt but cannot destroy these canny little critters.

THIS PAST WINTER, A VERY BITTER ONE, I WAS STUNNED TO SEE, ONE morning in January, that the temperature—real temperature, not wind chill—on the mountain was -20 degrees. As I filled the feeders for the hungry birds, gratified that I could at least help them survive such cold, by feeding them and thus helping them keep warm (as well as throwing out apples, corn, and peanuts for the turkeys, squirrels and deer, possums and porcupines, and any other opportunistic critter), I worried about the peepers, even though I knew that they could survive being completely frozen for several days. Sometimes, in the face of such extremity, when my face seems to freeze in seconds and my fingers, even in gloves, go numb as I fumble with the feeders, it is hard for me to believe in the smooth, sure assurances of science. How do any of these wild critters survive in these harsh Catskill winters? Even the coyotes sounded plaintive, their night wails pitiful, standing, I knew, shoulder deep in the snow, which was relentless, so that the ground was covered from December until well into March, the white drifts sheltering the small mammals the coyotes prey on.

The spring, this year a calendar assertion only, has been raw and wet, which the peepers love, even as the rest of us crave the sun, for the rains suffuse those pools and puddles that are these diminutive frogs' breeding grounds. I have heard the peepers, in milder winters, as early as the end of March, but it wasn't until the second week of April, as nighttime temperatures finally rose above freezing and into the forties, that I heard that first seductive whistle, the signal of desire of peeper males. That solitary high-pitched peep soon became the voices of many, almost as if that one amorous amphibian had awakened his bachelor mates, and their courting song seemed to ricochet off the mountain, suddenly so raucous after the stillness of winter, enveloping me as I stood there, out on the back deck in the chilly dusk. Sometimes, when spring comes so slowly here, bringing frost and snow showers instead of wildflowers into April, it is easy to wonder, senses trumping common sense, if it will ever come. That evening, after wrapping myself in my down jacket and donning my wool cap, scarf, and mittens, I went back outside, dressed for winter, where I sat down on the back deck steps and listened, just listened, well past dark, to the sound of the spring peepers, sleigh bells jingling on the mountainside, and the miracle they herald—the return of spring, of life, to the Catskills.

CHAPTER THREE

Summer of the Flood

(For Edward Abbey)

T HE WATER IN THE MOUNTAIN IS GENTLE, FILTERED BY ROCK, fed by springs replenished by rain. In spring, when the snow melts and is released as water, it seeps almost reluctantly. The winter mountain keeps its secrets, is, in some places, impassable, even unrecognizable. That frozen world, aloof, severe, punishing even, calls a halt to life, forces it to retreat, to seek the shelter of slumber—its symbol in the Catskills, the black bear dreaming in its den. Life is restored by water, by the warm spring rains that create vernal pools for the tiny tree frogs to breed, their "peeping" the first siren song of spring; that return the American robin, the cheerful thrush with the chestnut-red breast, to the mountain, freeing the soil for worms to migrate to the surface, the earth's first offering of food, transforming snow into mud, that modest though crucial commodity the robins, and many other birds, use to fashion nests.

The "mud season" here can last into June and "will suck the Wellies right off ya," according to Farmer McCagg, whose herd of Here-fords range along the West Bank of the Delaware in the valley below. Water is everywhere in the Western Catskills—rivers, streams, ponds, and though least conspicuous and largely uncelebrated, the springs that nourish and sustain them. The pond on my property is spring-fed, offer-ing a haven for green frogs and the largemouth bass that pursue them as well as cattails that grow at the water's edge, providing nest sites for red-winged blackbirds and reedy heights for dragonflies and damselflies to command. The pond, a calm, intimate body of water, is my ally too, the only source of water on the mountain in case of fire. Walking the land in spring, I can easily ferret out the springs—they are the stubborn patches of wet that are slowest to recede. I am also alerted to springs by water-loving wildflowers that colonize the moist surface. In May, sweet white violets, their diminutive blossom featuring five pale ivory petals—including a lower one with a purple branching pattern, a delicate, winsome signature—lead me to an area that rarely runs dry, even in summer.

Another spring is the source of my drinking water—hard, it has an undistinguished taste and is not particularly refreshing. Take the Cut into the woods, head up the stream I clear every spring, which starts as a halting flow of water emerging from the mountain. Just above it and to the right is the "spring house," which marks where another spring trickles underground, captured by piping that carries fresh water into the house, where it is filtered and purged of unwelcome organisms by an ultra-violet light, precautions that are amenities of the modern Catskill "settler." The term "spring house" is mostly anachronistic but gets to perhaps our most domestic relationship with wild water. A small "house" would sit over where the spring emerged to protect the water source from animals and detritus such as fallen leaves. It was also used for natural refrigeration to keep perishables, especially milk and meat, cool. "Spring house," today, is usually a designation of place—where the spring, ensnared in a reservoir, begins its journey, often marked by a bluestone slab that covers it. Even so, the idea of a "spring house" still resonates, offering the assurance that water can be managed, controlled, is our willing partner.

Everything depends on water—on the alchemy of life it creates with sunlight. In the Catskills, we can see that, touch it, every day. In prolonged periods of drought, which are not usual here, I check the stream to determine its flow from the mountain, which is a living organ-

ism as the water is, capable of altering the spring's direction, even oblit-
erating it. That has happened here, and the spring, which had ceased to
flow into the cistern, willfully escaping around it, had to be recaptured,
leaving me panicked, waterless for several days. Water, slippery, mercurial, always unpredictable—the ultimate shape-shifter—even in the guise
of a spring, that mild, amenable spirit, will always have the upper hand.

FOR THOUSANDS OF YEARS, THE LENNI LENAPE—THE "ORIGINAL
people, or men that are men" as their name is usually translated—
dwelled in lands now encompassed by New Jersey, Delaware, eastern
Pennsylvania, and southern New York. In the Western Catskills, the
Lenape clan was the Munsee (or Minsi), "the people of the stony country," who lived along the banks of the Delaware, both the east and west
branches, perhaps as far north as present-day Margaretville. The appel-
lation "Delaware Indians" was given to them by the first Europeans to
arrive in the 1600s, who named the Lenapes' river "Delaware" in honor
of Thomas West, third Baron De La Warr, the Virginia Colony's first
royal governor. Though not their own, the term "Delaware" to designate
these indigenous Catskill peoples was an accurate indication of the importance of the river in their life, a waterway flowing through Lenapehoking, "Land of the Lenape," into Delaware Bay, where it joined the
Atlantic Ocean. The river sustained them, offering the Lenape an abundance of fish, notably shad, and game: the elk, black bear and deer that
also frequented its waters, providing a ready supply of meat and hides.

  Early settlers, many from the British Isles as well as New Englan-
ders and emigrants from lower New York, came to the Western Catskills
in search of land. The village of Walton on the West Branch of the
Delaware was settled by a group of Long Islanders in 1785 who had been
lured by "surveyors' reports of a wide and fertile valley" with a river running through it that promised the same riches it had bestowed on the
Lenape. That river, the Delaware, would not only nurture the settlers
but also foster commerce; it became a thoroughfare for rafting local

commodities such as lumber and bluestone to the port of Philadelphia. The frequent flooding of its banks even offered the settlers the recompense of replenishing the soil with rich, alluvial till—though it was a reminder that the Delaware, a seemingly slow-moving, sulky river, was ever treacherous.

The Catskill settlers I represent are "Flatlanders"—the people of the Flatland, downstate New York and especially, New York City. It is a somewhat pejorative though occasionally affectionate term mountain folks use to describe city slickers, with a connotation of ineptitude, especially when it comes to coping with country life. In my own defense, as a native Philadelphian, I have an affinity for the Delaware, which flows by that city, though the wide, industrial river of my childhood, replete with tugboats and barges, bears little resemblance to the West Branch in the valley at the foot of the mountain, a rather unimposing conduit, home now to nesting pairs of bald eagles, which can easily pluck fish from its shallow waters. I also lay claim to the family lore— which I cannot verify but cling to—that I have the "blood of the Lenape" in my veins, thanks to a Scottish ancestor, a blacksmith, who settled in "Penn's Woods" on the Delaware River in the eighteenth century and who, in my father's words, "married well." Of course, this tale was told when I was acting, as my Dad used to say when I was little, "like a wild Indian" ("Arthur, please," my mother would scold, "do *not* encourage her!"), proof perhaps that it was apocryphal, as the Lenape were a peaceful people, revered for their skill as mediators, who lived in harmony with nature and the world around them until they were usurped by the end of the eighteenth century, banished forever from their ancestral lands.

Hapless Flatlander that I am, I discovered the Western Catskills, unlike my noble antecedents, not by design but by accident. I simply took a wrong turn. I had been upstate "leaf peeping," seeking the colors of autumn which in the city never seem to amount to more than a resigned brown, the leaves soon curled into their winter retreat. Reluctant to leave, I headed west, following roads I had never traveled, past dusty towns—more often, "villages" or "hamlets," sleepy enclaves consisting of several blocks—with unfamiliar, even exotic-sounding names like Fleischmanns and Arkville. Ulster County, before that trip, had formed the perimeter of my Catskill experience. The High Peaks of the Eastern Catskills, with their deep cloves and arid crags, sheer waterfalls and desolate wilds, had captured the imagination of Thomas Cole and other painters of the Hudson River School. But Delaware County, which I was now driving through, offered a different landscape, easier to apprehend, to make one's own: the Catskill Foothills, still imposing, but rounded, gentle, and so very green. I meandered alongside a sluggish river, which seemed to me more a stream, but which I learned at Margaretville, another frontier town, was the East Branch of the Delaware, the river that rambles through the county that gave it its name. The river was embraced by a spacious, sunny valley, the light—sharp, crisp, on this cool October day—was lucent, and fuller than the light of the High Peaks, whose steeper slopes and narrower valleys seemed to staunch the afternoon sun.

The East Branch soon widened into what I could only call an inland sea—it was vast, the light glancing off its endless waters—flanked at the shoreline by deep green pines, maples and aspens, beeches and basswoods, the deciduous trees torrid with color. The "sea" was, in fact, a reservoir, the Pepacton, its dimensions astonishing—seventeen miles long, with a shoreline of nearly fifty-one miles, nearly a mile across at its widest point. The Pepacton had been created in 1955 to slake the thirst of the Flatland—New York City—and not without controversy. Nine Catskill villages were razed, their buildings demolished, their people and even their dead, removed. At the time, seeing these signs posted on the roadside, I felt haunted though I didn't then know why: "Former site of Arena . . . of Pepacton . . . of Shavertown . . . of Union Grove . . ." Nature had been restored here by the waters of the East Branch of the Delaware River, a quarter of which had been impounded at the nearby

Downsville Dam to make a reservoir, flooding this valley, leaving no trace of its human history, overwhelming even memories.

Perhaps ironically, it was the captured waters of the Pepacton that first intrigued, then enticed me to the wild Catskills. In truth, it was their tranquility that attracted me, the smoothness of the water's surface, unruffled even on a breezy day. Pepacton derives from the Lenni-Lenape word *Peapackton*, meaning "marriage of the waters," suggesting a harmonious union that seemed evident that afternoon. The East Branch melded into the Pepacton, softly surrendering to it, and the streams that fed the river were just as amiable, playful but acquiescent, obedient to their rocky course, escorting me as I ambled along. Water was everywhere, glittering water of the deepest blue, tranquil lakes and snug ponds, and though sometimes a trickle, parched by the summer sun, lean brooks that persevered. I felt soothed by the waters of Delaware County, welcomed in, and most of all, I felt safe here.

I HAD, THE PREVIOUS AUGUST, SPENT EIGHT DAYS ON THE COLORADO River, rafting the entire length of the Grand Canyon, 277 miles. I had taken no photographs, had taken no notes. It was a landscape I could not encompass, in pictures or in words, even in conversation, with the aid of gestures and inflection. Its scale was too monumental, even, at times, terrifying. The rock walls of the inner gorge rose up on either side of us a mile high, as we journeyed downstream. I remember at mile 65, looking up and seeing the morning light firing the South Rim, its gray sandstone cliffs stained red from iron oxides leaching out of the rocks, the Desert View Watchtower barely discernible. The seventy-foot-tall circular stone structure—built in 1932 as homage to an ancient Anasazi watchtower, the work of the Ancestral Pueblo Peoples—is the most prominent architectural feature on the South Rim, affording visitors a panoramic view of over one hundred miles. But from where I was, it seemed no more than a thimble—and to any tourist perched on the South Rim, contemplating the ribbon of river below, as it bent west, snaking inexorably downstream, we would have appeared, in our river raft, as a cork bobbing on the Mighty Colorado. And that's the terrifying part, at least for me—the Grand Canyon is so vast, and in it, you realize your own utter insignificance.

The scale of time is just as intimidating. The oldest rocks at the Canyon's bottom—"Vishnu basement rocks"—are two billion years old. At mile 178, we encountered "Vulcan's Anvil," a large cone of black basalt thrusting up out of the river, the remnant of a volcanic plug. Lava erupting from volcanoes on the canyon rim over the past two million years had dammed the ancient river; several of these "lava walls" had reached a height hundreds of feet high, creating "reservoirs" that rivaled today's Lake Powell and Lake Mead. But the Colorado, over thousands of years, had eroded these, washed them away, except for this stubborn slab of volcanic rock, an eerie reminder of both the Canyon's antiquity and the river's strength. Then, as it was when John Wesley Powell, the one-armed Civil War veteran, first rode the river in 1869—sitting backward in a wooden chair lashed to a rowboat—it was an ocean, undammed. But the Colorado, though today a river controlled, is not restrained, and being on it—rafting its rapids and most of all, seeing the reflection of its power—had changed my relationship with water.

I simply could not grasp that this canyon had been carved by water—even over the course of millennia. It seemed so much easier to believe, sitting in camp one evening in the deepening dusk, which fell swiftly in the canyon because of its walled heights, looking up at the stars swirling in that chasm of sky, as western red bats swooped over us, that this rent in the rocks of the Grand Canyon had been made by the hard, swift hand of God. The wildlife we encountered on the sandbars where we camped—the bark scorpions, small but savage, that nested in a sleeping bag; the western diamondback "pink" rattlesnake sleeping in the afternoon sun; the chittering ring-tailed cats that swarmed out of the canyon walls at night, intent on stealing our sunbaked cakes—these did not frighten me. What left me uneasy and relieved, finally, to be back on land, was this throttled river, which, left unchecked, would over time erode even Glen Canyon Dam. Its nightly releases, traveling downriver, were like a sudden tide coming in, always waking me as I slept on the sandbar, worried that I, too, could be washed way.

I sought softer landscapes after that, one where I could perhaps still find, in places, a "howling wilderness," as David Brainard, a Christian missionary intent on "saving" the souls of the Lenni Lenape and other native peoples, had characterized the Catskills in 1744. But I also wanted a landscape—a vista—that wouldn't burn my eyes, as the relentless, arid rock of the Grand Canyon had. I wanted heights that wouldn't overwhelm me, force me to contemplate my own meaninglessness, and perhaps most of all, I wanted waters that would not roil, devoid of rushing rapids, rivers whose might would not summon, in my imagination, the power of the divine. The Western Catskills—its cascading green a solace to the senses, its bluestone eruptions a reminder more than a declaration of the antiquity of these eroded foothills, its cool, gray, misty days a consolation, and most of all, its crisscrossing blue waters, so clear after the Colorado in monsoon season, livid with the red mud that once gave that river its name—seemed the wild respite I was searching for.

I HAVE ALWAYS TAKEN COMFORT IN RAIN. IT IS THE MOST INTIMATE of waters. Nothing soothes like the sound of rain, lulling me to sleep. Rain seems to rouse the peepers, making them sing louder, even more joyously, as the wetlands where they breed are replenished. A fair rain, steady but falling gently, is what the farmers call a "good growing rain." It soaks the earth but doesn't wash away soil, seeping into the ground to nurture roots. It always amazes me how after such an even, insistent rain, especially in May and June when new life is at its most eager, the grass seems to grow inches, is so much thicker, luxuriant, sparkling with morning dew. Even the rockiest soil, stubborn, resistant to growth, will green as it is raining, overspreading the red clay with a luminous, almost-emerald green—short-lived, if it doesn't rain again soon, but at least briefly defying the Catskill dictum "two stones for every dirt."

Rain is welcome in the Catskills, even a relief. It means that we will have water, our springs and wells recharged, that the farmer will have hay for his cows, that his corn will grow high and green, a gracious crop. Even the smell of it—that sweet quick freshening—lifts me up, restores my spirits. I love watching rain in the valley below, a white billowing mist chasing the river. Sometimes it veers off, starts up the mountain. The rain ascends swiftly, the mist now a linear curtain, tendrils of gray. I can hear the *whoosh!* before it arrives, rain sweeping upward, until the first solitary drops strike the deck, resuming its familiar consoling patter.

The darkening sky and the rumblings that signify a thunderstorm's approach hold more thrill than fear for me. Even in the Catskills, where thunderstorms can be severe, the pelting rain forcing rivers and streams out of their banks, muddy waters swirling over roads and pastureland—a sudden flash sparking a fire, a hurled bolt blasting a favorite tree, lightning in its scintillating, wild beauty—there is the assurance that they will end. Thunderstorms are an ordered part of nature, they have a consistency, predictability, even—a cause and effect we can count on. But the summer of the flood, the rains that came would not stop. They were so relentless, so utterly without remorse, it was almost as if they had a will of their own.

THE BIRDS STOPPED SINGING. THEIR CALLING, SUCH A COMPANIONABLE part of a June day in the Catskills, ceased. Birds have a sixth sense about rain. I can time when rain will arrive, even when the thunderstorm that has been grumbling in the distance will begin, to when the birds fall silent. They also tell me when rain is about to finish. One brave bird— a song sparrow perhaps, its bright clattering notes a fit tribute to the storm's end—will start to sing, even though it is still pouring. When the rain began to fall on Monday, June 26, 2006, I expected their silence. But even when the rain stilled, offering brief respites, the birds stayed quiet,

almost as if they didn't want to reveal where they were to a predator—
this rapacious rain.

The ground was already saturated. Everywhere I walked, mud
sucked at my boots. June had been wet in the Catskills—several storms
had crossed the region, each leaving as much as one and a half inches of
rain. We need rain here in June, the heart of the growing season, but by
mid-month we had had enough. Still, the rain persisted, the streams and
creeks running high. Sun and rain require a balance—too little of one,
too much of the other, crops are lost. It is a balance that is fluid, that can
sustain some disruption. The summer sun works fast, even in the usually
cool Catskills, drying out the soil. So we wish for rain, embrace it, watch
the skies for that reassuring seam of gray that means rain is on the way.

On Friday, June 23rd, guests arrived for the weekend to rain, light
but persistent, lasting through the afternoon. I remember apologizing for
asking them to take off their shoes in the house—Catskill etiquette—and
for the mud season "lasting so long this year." "I guess it'll be July before
we're dry," I had laughed. Then, on Sunday, June 25th, the rain picked
up, registering over two inches in the gauge before we went out to dinner.
The roads were clear, I was reassured to see, troubled only by scattered
puddles. But by the time we returned, it was difficult to get up the mountain, the dirt road having metastasized into mud, the car fishtailing even
in second gear. We escaped into the house. More rain was due tomorrow,
another one to three inches. The flatlanders fled early Monday morning,
thanking me for their "Catskill experience," before the rains resumed.
By Tuesday evening, June 27th, the roads they had taken were closed,
Delaware County was in a State of Emergency, and water was rushing
through the streets of Walton, the village in the valley on the West Bank
of the Delaware, and would continue to rise through Wednesday.

Monday night, the rain continued—a hard, pounding rain—and
for the first time, I felt apprehensive. Perhaps it was the cats who, like
most felines, are wild at heart and understand things we don't. They
stayed huddled under the bed and refused to emerge, even the next

morning, to eat. When it wasn't raining on Tuesday, I was relieved, though the cats were inconsolable, crouched now under the covers, where I was forced to feed them. The day was eerie—too warm for June in the Catskills, oppressively humid, steam rising from the decks, the sun an orange mask, the sky bleached white. It was also quiet, strangely quiet. The birds weren't singing, I realized. I searched the sky but could spot only a few crows flying in the distance. Even the ruby-throated hummingbird, the "little king," who loved the rain, preening himself in thunderstorms, was not at his post, guarding the sugar water feeder.

I checked the rain gauge and found that three more inches had fallen since yesterday, and the forecast was for more rain Tuesday afternoon. But living on the side of a mountain, I felt fairly confident. The West Branch was in the valley below, and though I had seen it rise, its waters overreaching its banks, even spilling onto the main street of Walton, they had quickly retreated. My own waters were small and docile. Wee Brook, which drained the pond, preventing it from overflowing, trundled down the mountain far from the house, finally flowing into a stream that fed the West Branch. Little Porcupine Creek, in the woods, which I had carved out of a bog and cleared every spring to prevent it from being overtaken by weeds, was flowing well—a good sign for the spring house, which channeled the same waters—but was within its low banks. Slogging home through the wet, glad for my Wellies, I saw a mallard hen and her eight ducklings, following her closely, hurry by. They were too intent on their purpose to notice me. Then, even more surreal, a snapping turtle lumbered past, in the same direction. The mallards, the snapping turtle—they were not residents of my small pond. They must have come from higher up where, I knew, several large ponds offered more spacious accommodations. Where were they going? I wondered. And what did they know that I didn't?

I LOST POWER AT 10:00 PM THAT NIGHT. IN THE DARK, THE RAIN WAS even more terrifying. I stood at my bedroom window, straining to see.

But everything was black. The night, the rain's furtive ally, played on my fears and amplified, in my imagination, that hideous sound. There was no comfort in this rain. It had a might I had never before experienced. The rain was dangerous in the way that only nature is—a relentless, unforgiving force that cannot be stopped until it extinguishes itself. This was the nature I had glimpsed in the Grand Canyon, which had fascinated, even enthralled me, but which I had been relieved to leave behind—the nature that is so much bigger than we are, so much stronger. The power of water, its effects that I had witnessed there, returned to me as I listened to the roaring in the woods. Little Porcupine Creek, no longer tame, sounded as if it was moving closer, closer . . .

I was too frightened to sleep; hearing the rain, the only word I could think of to describe it was "biblical." Instead, I found myself praying silently, then as the rain wouldn't cease, out loud: "Please, God, let it stop—please, let the rain stop."

I STARTED AWAKE ON WEDNESDAY MORNING, MY HEART POUNDING. IN spite of myself, I had ceded to sleep. Morning had come—no comfort to the cats, who were hiding, curled up in the closet. It was still raining, but lightly, an opaque mist rising over the meadow. Finally I heard it: birdsong, the signal the storm had passed. It was a robin, that sturdy songster, celebrating his second nest of the season, as if all was right with the world. Then I heard a new sound: helicopters, two of them, flying low over the house, heading for the valley. From the living room window I watched as they disappeared into the mist, which had re-

ceded enough to reveal the muddy red torrent that was the West Branch, as wide as the Colorado I had traveled, slashing through the valley below. The helicopters, I would learn, were plucking people off of rooftops. Walton was underwater, as much as eight feet of it.

At first I thought it was an optical illusion. From the kitchen window I could see what looked like a wave of water undulating down the road, like some great mythic sea monster worthy of Homer's *Odyssey*. At the end of the dirt driveway, a gorge some three feet wide and deep, where water had erupted from a drainage ditch, had separated the driveway from the road. At least I could jump over it. But on the other side of the road, I was shocked to discover the monster was real. The stream had jumped its banks due to a slide of rocks and mud and a sluice pipe clogged with tree branches, pushed by more than nine inches of rain that had fallen overnight, judging by my gauge. Water had poured forth with such force that a rift, over ten feet wide and deep, ran along that side. I escorted the wave down the road, carefully edging this new Catskill clove that had opened so precipitously, until the water jumped again, following its own mercurial logic, this time returning to the shallow ravine that carried the still-boiling stream to the West Branch.

I kept walking, wanting to check on the condition of the road, which had been heavily gouged by the rain. Where the road bends, starting its final descent to the valley floor, I paused, trying to fathom what I was seeing. Another creek had gone rogue, charging down the rutted side path that led to the bluestone quarry, then taking a sharp turn onto the road where I was standing. Its journey had been violent, brutal even—the dirt road, yesterday about eighteen feet wide, now measured no more than six feet across, barely enough for a car to creep by. The road was just gone. In its stead was a canyon—its crumbling walls steep, a dangerous drop-off— some twenty-five feet deep. Water had done this, in a few hours . . .

MUD WAS EVERYWHERE. EVEN THE TREES—THOSE THAT HAD SUR- vived the flood's onslaught—were covered in mud. A thick red dust hung in the air, dried mud kicked up by passing cars.

When I was finally able to drive into the village of Walton, several days later, I pulled into the rubble of what had been the parking lot of T.A.'s, the rustic diner that now listed precariously over Third Brook, its interior gutted, a side wall and half of its foundation gone, and cried. The spiteful waters had left behind two feet of sludge in Village businesses—Country Emporium, Poppa's Diner, Tony's Shoe Store—filling the basements of homes, as well as garages and gazebos, with mud. Asphalt had been peeled away carelessly, washed into the river. The sidewalk, now an eruption of bluestone slabs, was, in places, impassable. Concrete planters, filled with petunias, a recent attempt at beautification, had been sent "bob-bob-bobbing" into the Delaware. More mud, an ugly reminder of the flood's high-water mark, stained the sides of buildings. So many storefronts had already been abandoned in this struggling town. Now those that did have tenants, their businesses were destroyed. All the bridges were out. The village had been cut in two, cut off from the world.

Walton is crisscrossed by creeks—eight of them—as well as by the West Branch of the Delaware. The floodwaters had uprooted trees, sent them crashing into bridges, along with boulders and bits of buildings, forcing the creeks, suddenly turned savage, out of their banks and into the streets, where they surged through town with a singular ferocity. High, lopsided piles in front of house after house told the story, and revealed the toll the flood had taken: furniture, refrigerators, files, clothes, children's toys, board games, paper goods, bread and frozen meat—once the possessions, now the detritus, of people's lives. Some buildings had been so damaged by the rising floodwaters, they would have to be condemned. In the pharmacy, which managed to stay open at least to fill prescriptions—giant barn fans running on generators drying out the store, its soggy carpet ripped out, discarded in a heap along with Kotex and candy—I heard a woman tell a friend: "We lost our hot water heater, our boiler, most of our foundation, and all of our appliances. It could have been worse . . ."

A stalled frontal system caused tropical moisture to be fun-
neled northward into New York, causing severe flooding

in the Mohawk, Delaware, and Susquehanna River basins during June 26–29, 2006. Rainfall totals for this multi-day event ranged from 2 to 3 inches to greater than 13 inches in southern New York. [Walton recorded as much as 15 inches of rain, including the one-day high of 9.63 inches on June 28. The "beast," as the National Weather Service described the storm, had hunkered down over the town, refusing to leave.] The storm and flooding claimed four lives in New York, destroyed or damaged thousands of homes and businesses, and closed hundreds of roads and highways. Thousands of people evacuated their homes as floodwaters reached new record elevations at many locations within the three basins. Twelve New York counties were declared Federal disaster areas, more than 15,500 residents applied for disaster assistance, and millions of dollars in damages resulted from the flooding.

Chief Jim Jacob of the Walton Fire Department put the Flood of '06—already being called the "300-year flood"—in pithy perspective: "We had a flood back here in 1996, and this makes that look like it was just a day of rain."

But the sight that moved me the most, that changed my life in the Catskills, was the library, specifically, the sight of all those books, hundreds of them, water-logged, ruined, lying out front, littering the lawn. The ideas, learning and lore, the worlds of imagination were to be taken to the dump, the flood's forgotten victims. The William B. Ogden Free Library, a solid one-and-a-half-story Romanesque structure built out of local bluestone in 1897, had survived the flood intact. But the library's lower level had been inundated. Several important collections had been lost, including documents relating to local history, irreplaceable; children's books, an essential resource for a town with so many youngsters where books were a luxury; and the ever-popular computers, connecting Walton, as one high school student told me, to the universe. Libraries in rural areas are still important, a vital center for education and exploration, the heart of the community.

I joined that community, played my small part in the recovery efforts, begging and buying and borrowing books—I would have stolen them if I could—to help rebuild the library's ravaged collections. It was largely that spirit of cooperation, of neighbor helping neighbor, that in time would bring Walton back. "Pay me when you can," became a mantra. Poppa's Diner on Delaware Street—"Main Street," as locals call it—with its fifties décor, where Elvis was still king, reopened on July 5th, barely a week after the flood. "We got together and dug and we dug, so much mud you wouldn't believe it," said a waitress, "but we did it."

Before I came back up the mountain, after that first trip into town, I saw on a grimy window of a store that at least still had its windows a message scrawled in soap, elegant in its simplicity and defiance, its command of hope: *"Don't give up!"*

LITTLE PORCUPINE CREEK WAS STILL ROARING. THE DAY AFTER THE rains ended, that Thursday, I walked into the woods to see for myself. I found the creek funneling down the gradual incline where it used to trickle, especially as summer set in. Now it was leaping, looking like a watery slinky descending stairs, but it was still within its banks. That's because the banks had been widened, deepened. This was a spring-fed creek, the water would soon calm, become once again dutiful, as it drained through the mountain, finding its way into the West Branch of the Delaware. But my spring house was also part of this system. Eventually, that would have to be rechanneled, its waters captured anew, as the underground spring, driven by too much rain, would seek a new course. The Lenni Lenape believed that everything in nature is alive, has spirits that dwell within. One only has to live in the Catskills a short time to understand this, to feel the companionability of all things, even the springs and the rock that makes up the mountains they inhabit.

That afternoon, as I was checking the outbuildings for damage, I saw the mallard mother returning with her eight ducklings, walking less urgently now, heading back up the mountain. I also spotted the snapping turtle, that prehistoric survivor, making its own slow, laborious trek back to where it had descended to safety from the rain. I had wondered, the day of the flood, why they were leaving higher ground. It seemed counterintuitive. But now I realized they were fleeing the rising waters that would have overwhelmed them—the very ponds where they lived. They were the lucky ones. Many wild and domesticated critters, too, are lost to storms such as this, washed out of trees, drowned in raging rivers, the silent victims of natural catastrophes, fire and floods, especially. I was glad to see these incongruous sojourners, and especially glad that the mallards were on a different path, as those ducklings, even Mama herself, might have proved too tempting for the voracious snapping turtle, which, though it appears lethargic, can strike with lethal swiftness, grasping prey in its powerful primitive jaws.

THE NEXT DAY, I WAS DOWN BY THE ROAD, WRESTLING WITH MY MAILbox, which the floodwaters had knocked down. I was trying to lean it against the gate until I could have it reset so that I would be able to get mail when the road was repaired, enabling the mail truck to venture up it. "Well, I'd offer to help," said the familiar voice, "but if you're going to live in these hills, it's best you learn how to be self-sufficient." I turned around and had to laugh. It was Edna, my neighbor who lived in the valley below. Her people had been here for generations—had farmed everything from cauliflower to dairy cattle—she still lived in the old farmhouse, which was safely situated up on a knoll. She was in her seventies but hiked up the steep mountain road every day, her "constitutional," she called it.

"I appreciate the sentiment," I said, pausing in my struggle with the mailbox, "Long time, no see, Edna."

"We had a little shower, I hear," she continued.

"Just a little one," I said, and we both laughed.

"That's a pretty good pothole you got back there," she said, indicating the road with a jerk of her thumb. Then, after a pause: "You know, Flatland," she began, her nickname for me, which, I presumed, was affectionate, "I'm kinda surprised to see you here. I woulda thought you'd be *long* gone by now."

"Nope, I'm still here," I told her, "and I ain't goin' nowhere."

CHAPTER FOUR

Natural Beauties

(For Loren Eiseley)

F INALLY IT WAS WARM, AN EVENING IN EARLY JUNE. THE FIRE-flies had begun to rise in the meadows, twinkling in the high grass, signaling their desire. I could hear through the window—opened at last after a chilly spring—the plunking of green frogs at the pond, and, though their chorus had diminished as summer approached, the insistent whistling of spring peepers, stalwart suitors reluctant to cease their courtship song. The night air was redolent with the sweet, delicate fragrance of wild rose, my thorny adversary, tough and recalcitrant, suddenly turned ethereal, its white blossoms covering the mountain like a spring snowstorm, gentling even the harsh, stony Catskills.

Stepping outside shortly after midnight to answer the insistent calling of a barred owl (*"who-cooks-for-yoooouuuu"*), I noticed how moths were circling the security light, perched on a metal pole set several yards from the house. I had been advised, when I first moved to the mountain, to keep a light on at night to deter critters from climbing onto the back deck. The bears had behaved, but porcupines especially had not been deterred, routinely gnawing on the wooden decking, causing several boards to be replaced. Opening the back door one morning, I had discovered a young porcupine small enough to curl up on the coir mat where it had fallen asleep, which I gently rousted with a broom.

I was mesmerized by the moths, swirling in the milky, vaporous light. At the edges of the luminous halo dark shadows appeared—bats, nemesis of night-flying insects, hunting prey.

Suddenly they plunged into the light, scattering the moths, which despite the onslaught, would not abandon the glowing circle. The bats were also drawn to the porch lights. On summer mornings, I often found shards of moths, fragments of wings mostly, on both decks. The bats, I quickly learned, would devour the moths' succulent body but leave behind their pesky appendages. The porch lights have their own ecosystem; the lights attract moths and in turn their predators, the spiders that spin their messy webs, waiting for the moths to alight, and the bats, which feast on the insects that assemble there. I felt no remorse for the moths— such drab insects, small and flitting, in barely distinguishable shades of brown and gray, many of them notorious pests, at least in their voracious larval form, caterpillars.

The night-flying adults mostly go unnoticed and those that settle on screens on a summer night, seeking light, are usually ignored, except by cats, which know them as the fascinating creatures they are. Cats are the ideal naturalists, patient, persistent, ever watchful—their powers of observation are unsurpassed. My own cats, a tri-colored calico and a pearly gray, alert me to the comings and goings of critters, especially birds alighting on the deck rail: a gray phoebe flycatcher twitching her tail; young robins, newly fledged, sitting in a row, fuzzy and speckled, solemn after taking their first flight. The cats chatter their staccato hunting call, indignant at being ignored by their prey. There are other nosey visitors—gray squirrels, chipmunks, and white-footed mice—which tantalize the cats, coming so close, peering at them through the windows. The little red squirrel, with its devilish ear tufts, brash and irreverent, seems to enjoy mocking them, racing up and down the front deck, leading the housebound felines on a futile chase. If a moth manages to slip in, they pursue it recklessly, leaping into the air, clattering over tables, knocking everything off until I manage to corral the hapless insect and escort it outside.

The calico, who is also the clown, stubby and graceless though very sweet, often finds herself in need of rescuing from her misadventures. It wasn't the first time I would find her hanging by her claws from the screen door when I came in that night after watching the moths and

bats dance their skittering minuet. Usually she would cry, unable to extricate herself from the screen, a mournful keening that crescendos into a yowl. But this time, when I went to rescue her, she was chattering, her eyes fixed upward in that intense, narrow concentration so characteristic of cats, pointing at her prey. When I saw what she was stalking, I stepped back from the front door, and gasped . . .

The creature was huge, at least compared to the other moths clinging to the screen, lured by the porch light. It had two long tails and wide, arcing wings that spanned over four inches. But what captivated me was its color, the tender green of the spring forest, newly leafed—a green that had an odd luminosity, as if it glowed from within, like an egg tempera painting. I reached out to touch the creature through the screen—it fluttered off, disappearing into the darkness. "Tinkerbell," was my silly, thoroughly unscientific assessment. It was the first time I had ever seen a Luna moth.

There are some creatures that are so beautiful, they don't seem real. Their colors are so vibrant, their markings so crisp, they look painted. Birds especially, including Catskill summer residents such as the wood duck, cedar waxwing, and many wood-warblers, with their variegated and often spectacular plumage, have captivated artists from John James Audubon to Roger Torey Peterson to the contemporary master

David Allen Sibley, whose finely etched watercolor illustrations convey the delicacy and subtlety as well as the boldness and brightness of their avian subjects.

But just as striking, if less well known and certainly underappreciated, are Nature's other beauties—insects. Butterflies, of course, ever showy, get most of the kudos. They are first eminently visible, flying by day, flashing their vivid colors, revealing often intricate patterning as they alight on flowers to suck nectar through their proboscis, a long, thin feeding tube that is coiled in flight. Their colorful wings, catching the sun, help hide them from predators, making them difficult to see among the blooms—camouflage that, for us, is eye-catching. Their color may also enable them to identify and communicate with each other, helping butterflies to find a mate. In some species, a butterfly's colors can signal danger. The monarch, as regal as its name, is colored orange and black (colors whose resonance of "danger" we have inherited in Halloween), a signal for predators such as birds to stay away. They taste awful, the result of the monarch laying its eggs on the noxious milkweed, and another species, the smaller viceroy, mimics the monarch's coloration and though not bitter tasting, benefits from its bad reputation. The monarch also has romance—its extraordinary cycle of migration, a fall journey of as much as 3,000 miles, as monarchs in North America travel from as far away as Canada to roost in sites in California and Mexico, where they overwinter, and then return to breed the following spring. Even butterflies' names are stirring and magical: the monarch and its vassal viceroy, painted ladies and red admirals, swallowtails (tiger, spicebush and the elegant, formally attired black swallowtail), satyrs and clouded sulphurs, question marks, commas and the mourning cloak, the great spangled fritillary and the diminutive pearl crescent, which always seem to hatch after a Catskill thunderstorm, to name a common few of North America's approximately 750 species.

Second only to the Coleoptera (better known as beetles, the largest order in the animal kingdom), 150,000 species of Lepidoptera—

butterflies and moths—have been described (or formally identified) worldwide. Of these, only 20,000 are butterflies. The vast majority, some 130,000 species, is largely ignored, at least until they plague us: moths. To be sure, many moths deserve our opprobrium. We encounter them often as destructive insects—the larvae, in caterpillar form, feeding on wool sweaters and devouring grain, miscreants known as clothes moths and pantry moths, respectively. Their rural cousins, forest tent caterpillars, can decimate the leaves of some of our most beautiful and economically important trees in the Catskills, notably, sugar maple, oak, aspen, apple, alder, elm, cherry, and birch.

But far more grievous is the destruction of habitat—for which Nature has no remediation—at human hands. Frequently, it is the butterfly, especially the monarch, which is the poster bug for this loss. Its host, specifically for its larval stage, the milkweed—an invasive though fragrant native species, whose pinkish blooms are intoxicating in their sweetness—is rapidly disappearing in the dense stands the monarch needs to successfully rear succeeding generations. The monarch's beauty is an argument for its preservation—its bright orange wings boldly veined and bordered in black are iconic, as is the monarch's immense journey, a marvel, and mystery of Nature perhaps only equaled by the migration of the fragile hummingbird, weighing less than an ounce, which also braves a treacherous two-way trip, traveling from the Catskills and points even further north in fall to destinations as far south as Central America, then returning "home" to nest the following spring.

"Don't there seem to be fewer butterflies this year?" is a refrain I hear often in the Catskills. Such anecdotal evidence may well be true, an indication of the impact of pesticides as well as shrinking wildflower fields on insect populations. But the point is that the butterfly, with its vivid colors, fluttering from flower to flower on a sunlit summer day, is readily missed, its loss lamented. Every schoolchild learns that butterflies are not just pretty but also useful pollinators. So is the moth, which, like the butterfly, appeared on Earth tens of millions of years ago, with the advent of flowering plants. Yet this fellow Lepidoptera is mostly ignored and stigmatized for the sins of a relative few (and butterfly larvae, though not as notorious, are just as gluttonous as their moth brethren). Such is the prejudice of beauty, how it biases us, which I am the first to admit I share.

It wasn't until my encounter with *Actias luna*—the Luna moth—that I considered moths as anything more than prey, a favorite fare of bats and birds, spiders and praying mantises, amphibians such as frogs, toads and salamanders, and the terrestrial mammals, too. I often watched, with clinical detachment, as skunks plucked moths off of the lower part of the light pole where they had landed. The opossum, shy and slow, would also visit, braving the brightness of the security light to scrounge a snack, as well as the clever raccoons, which scooped up their quarry with agile hands. That moths were so vulnerable didn't concern me until I glanced out the window later that night and spotted the Luna clinging to the metal pole, keeping company with the smaller moths, which surrounded it like courtiers. The Luna moth, a native species, is not considered rare in its North American range, though it is elusive, flying after midnight, its life span barely a week. In the Catskills, I have observed only a single generation of Luna moths, emerging from their cocoons to fly in June when it is finally warm.

"If a Luna moth lands on your hand, that's good luck," one Catskill old-timer told me. "Don't see 'em around too much here though."

I watched for a while, willing the Luna to fly off the pole, as the bats shape-shifted ominously through the circle of light. It dismayed me that something so beautiful should be prey, even though I knew how ridiculous, how utterly irrelevant, that notion was. Nature's beauty, if it acknowledged it, is in the adaptations it contrives for critters to survive and successfully reproduce. It plays no favorites; the dowdy moth, as commonly seen as the Luna is rarely sighted, is just as useful as pollinator and prey. Both, from Nature's perspective, are far more utilitarian than I am—a human, less ecologically significant than a moth, and far less important than an earthworm, which admirably discharges its duties, crucial to the survival of plants as well as animals, aerating the soil, breaking down organic material, and feeding generations of birds, toads, turtles, snakes, insects, mice and other small mammals (and its castings, or waste, even make a valuable fertilizer). Surely Nature would regard the lowly earthworm, unappetizing to our taste and "icky" in appearance and to the touch, as any child can attest, as "beautiful," even sublime, in both its form and function. But I am swayed by beauty, seduced by its aesthetic, even though I realize how false, how subjective a value that is, and as the Luna moth finally flew away, the green of its luminous pistachio wings flickering in the veil of the security light, safe for this night at least from the bats' coarse desires, I was relieved.

I LOOKED FOR THE LUNA MOTH THE NEXT NIGHT AND SEVERAL NIGHTS after, but that was the only sighting that summer. But in searching for it on the screens (usurping the cats, whose proper province this was), I gained entry, however glancing, into the secret world of moths and really saw them for the first time. The moths, supposedly predictable in their plainness, were also creatures of surprise, the reward for my scrutiny. One night, while examining a moth on the screen door, its wingspan some three inches, I was startled, when it flew off at my touch, to see a flash of red. That moth, the sweetheart underwing, has mottled tan-gray forewings, but its hindwings feature a pinkish-orange color. When the moth flies it reveals that reddish hue, which is meant to disorient predators. Both butterflies and moths have two pairs of wings, but butterflies at rest hold their wings vertically, over their body, touching, while moths mostly fold their wings flat over each other, so that the forewings obscure part of the hindwings. This variation helps the sweetheart underwing surprise its enemies and escape, and became for me a sort of metaphor for the moth, that its beauty is often hidden, especially the beauty of its adaptations. The virgin tiger moth, also a nightly visitor, its forewings black, veined in yellow, is striking, but it is the moth's deep pink hindwings, when exposed, that act as a diversion. The moth's bright color is also a signal of its distastefulness, a reminder for birds not to come back for seconds.

If I had met the rosy maple moth before the Luna, it might have been the one to attract me. Though not as imposing in size (the wingspan, at most, is about two inches), it is a startlingly pretty moth with its fluffy vivid yellow body and pink forewings, which also feature a triangular yellow band across the middle. Its bright colors make the moth visible and therefore vulnerable—a seemingly strange adaptation. The conventional wisdom is that these are warning colors, indicating that the moth is distasteful, though several birds including blue jays, chickadees, and tufted titmice are known to eat them. Naturalist Jim McCormac sees an additional, slyer purpose for what he calls this "colorful camouflage." Rosy maple caterpillars primarily eat maple leaves and as adults, are often found in close proximity to these trees. This moth, I noted, looked much as red maple leaves can look in fall, with their variegated color palette, and indeed the red maple is a favorite host. McCormac observes that the moth, when placed among red maple samaras (the helicopter-like fruits of the red maple tree), which are colored pink, green, and yellow—a stage it reaches just as rosy maple moths are hatching—is virtually invisible, providing safe roosts for the rosy maple during the day.

As brightly colored as the rosy maple is, the Pandora sphinx moth is cryptic, subdued. When I first saw a Pandora sphinx, I thought it was a dead leaf that had blown onto the window screen. Its patchy greenish-gray coloration, in hues ranging from olive drab to hunter, is persuasive camouflage—a moth wearing Army fatigues—and its elongated, notched wings (which can span four inches) reminded me of oak leaves in autumn, curled and dry, which makes it easy for a predator to over-look, mistaking it for litter. Flying at night, many moths wear suits in shades of gray and brown, allowing them to melt into the darkness and evade their enemies. Such somber hues can also help moths retain their body heat, keeping them warm.

These discoveries were made at night. But, as I would learn, a good time to "moth" is during the day. I became aware of moths cling-ing to doorjambs, hiding beneath deck rails, secreting themselves in small, dark spaces, sleeping away the day, a universe of critters I had never noticed. They would only rouse themselves if the sun found them and then they would fly off to find a new shady shelter. Their muted colors, I realized, also served them well in the daylight, helping them blend not only into manmade surfaces but also on tree bark and branches. But often the moths would hide in plain sight, a vantage out of the glare of the sun their only criterion for selecting a resting place. Even so, I was stunned when I saw it, a puddle of brown pooling in a corner of the back deck the morning sun had not yet breached, and I felt a frisson of fear; there is something so visceral in our immediate re-sponse to strange insects.

Anyone who has ever chased a butterfly knows what a futile task that is. Just as you come close, it flutters off and alights on a flower out of reach. Butterflies seem to have a sixth sense about humans who, as adversaries, are far less nimble than birds and dragonflies, among the swiftest, most agile of insects. As we bumble along in pursuit, butterflies, ever vigilant, survey us with their two large compound eyes, which are composed of hundreds of lenses, each in itself a tiny eye, each seeing an image that when taken together, makes a mosaic—a virtual wide-angle composite picture of its world. Butterfly eyes are adapted for color vision to see in bright light so they can spot not only their own species but also flowers whose nectar they covet, whereas the compound eyes of moths, which are dark to absorb as much light as possible, are better suited to see at night. Moths are ineffably drawn to light as well, as it is believed they fly by orienting themselves to the moon and stars, an infatuation that accounts for their passionate embrace of outdoor lighting.

Whether confused by artificial light or perhaps dozing off, deciding it is day, moths are easier to approach, fleeing only the rude curiosity of cats. During daylight hours, they are even more amenable, falling into a deep slumber, which is why a clever hiding place is key to their survival. This particular moth, though hiding in plain sight, was in deep shade, its wings and body wedged partly under two deck posts, which made it a difficult target. Its chocolate wings, which spanned over six inches, could have been a splash of mud on the deck's deep russet stain. I knew, instinctively, that it was special and probably rare, and while the Luna would become the anticipated companion of warm June nights, its loping flight and green iridescence summer's finest gift, this sighting would be the first and only time I would glimpse the Cecropia, the largest moth in North America.

"Majestic," "regal" even, are the words that come to mind when gazing at the Cecropia, so it is fitting that its name derives from "Cecrops," the legendary king of ancient Athens. On closer inspection of the sleeping Cecropia, I saw that its rich brown wings had a grayish cast, flecked in white, which gave the moth a frosted appearance. The Cecropia, lying in the shadows on the dark deck, was well disguised. But the moth, tinged in gray, would also blend into rocky surfaces, disappearing into the brackish colors of a Catskills bluestone boulder. The wings, both forewings and hindwings, were bisected by a rusty red stripe edged in white. The body, I saw, was also red, with white bands, and the areas adjacent to the abdomen were red as well—red, a warning color in nature, would serve to startle an adversary when the Cecropia took flight. These colors—brown and gray and flashing red—I had observed in other moths, strategies to help them hide and evade their enemies. But the Cecropia also featured prominent dark spots at the tip of each lavender-hued forewing— "eyespots," or ocelli, an ingenious device of some butterflies and moths to divert a predator's attention away from its body and that may also aid in mate identification. (This explains why we often see Lepidoptera with ragged wings, which, at least in part, can be sacrificed while the body cannot.) There were also several smaller dark spots hemming the forewings, more discreet but still diversionary targets.

The Cecropia is a big, handsome, brawny moth, without the delicacy and lightness of the Luna, in form as well as color, but they do share a striking characteristic. Each has on both pairs of wings other small eyespots, which are almost transparent. The Cecropia's especially are crescent-shaped, windows of red with a whitish center. The ocelli on the Luna's forewings are also crescents, while on the hindwings circles contain what appear to be quarter moons. For both, these eyespots serve a familiar purpose—to startle predators as the moth makes its escape. But these particular markings also indicate that both moths belong to the same family, the Saturniidae, which gets its name from the fact that several of its members, including the Luna and the Cecropia, feature eyespots that contain concentric rings that are reminiscent of the planet Saturn. The Giant Silk Moths, which the Saturniids include, are the largest moths in North America as well as some of the most brightly colored (though some members of the family are relatively small, such as the rosy maple moth).

Despite their impressive size (and that, in itself, may be a deterrent to predators), the Cecropia and the Luna are not dreaded pests—they simply are not abundant enough as caterpillars to defoliate the trees that host them. As is true of all Saturniids, the adults do not eat, drawing sustenance from food stored when the moths were in their larval stage, and they have only vestigial mouth parts, which accounts for their lifespan of only a week or two, spent seeking a mate. That is their sole reason for being, the reproduction of species—their extraordinary beauty, from our perspective, is a by-product of nature's adaptive purposes. Even the delicate, feathery feelers that distinguish many moths from butterflies are used to help locate a mate. The female releases pheromones that her suitors, flying in the dark, can "catch" using their antennae, even several miles away.

The Saturniids lack tympana; they do not hear (which is not characteristic of all moths, many of which use their hearing as a defense, a warning of a predator's approach). Which explains why, in succeeding years, I would often spot Lunas at night flying close to the ground, or even resting in the grass. My first response was to rush outside with my flashlight to see if the moth was injured, which it didn't seem to be, stubbornly refusing my well-meaning but misguided urgings for it to fly. The Luna was content to stay there and I couldn't figure out why. This made the moth, to my conventional, un-Lepidopteran way of thinking, more defenseless. But, as I would learn, this was also a strategy to counter predators, especially its nemesis the bat. The Luna can't hear the bat, which squeaks as it hunts, using echolocation, a kind of natural sonar in which bats emit cries, then listen to their echoes, which enable them to locate flying insects. If the Luna isn't flying, the bat, which does not depend on its eyesight to hunt (contrary to myth, bats aren't blind but their "sonar" gives them an edge in total darkness), cannot locate it as readily. My assumptions had been countered once again by nature's common sense, its elegant strategies for survival, seeking balance between predator and prey.

Another mystery I pondered is the Luna's coloration, its singular green, certainly unusual in the fashion world of moths, which mostly dress conservatively, depending on bold accents to add color. In its larval stage, this green makes sense—as a caterpillar, the Luna is the same pale green as its parent, the color serving as camouflage as it dines on the tender new leaves of red maple, beech and birch, and black walnut, some of the local trees it favors. As it prepares to pupate—and in the Catskills' chilly climate, overwinter, after a single generation in June (at least that I have observed)—the caterpillar turns brown, falls to the ground (where it wraps itself in a cocoon of its own silk, incorporating a leaf in its disguise), then hides in the litter of the forest floor where it will emerge as a moth when the weather warms. But what purpose could the adult Luna's shimmering green serve? Wouldn't it expose the moth to danger, flying at night and especially during the day? The answer to the latter question was answered during a June walk, when I almost stepped on a Luna moth, which fluttered up at my approaching footfall. The sleeping Luna, choosing the lush spring grass to nap in, was virtually invisible. The answer to the former question, how does its color serve the Luna at night, is perhaps that it isn't meant to. As I would discover, my sighting of the Luna was an accident caused by my intrusion into the moth's world—the bright, glaring lights that lure it away from the enveloping darkness and the relative safety of its vantage, resting or flying low to the ground.

THE LUNAS, AND THE SATURNIIDS ESPECIALLY, ARE BEDAZZLED BY light. It is a romance that distracts the Luna from mating and in the case of the female, from laying eggs. The fact that it is so short-lived, surviving about a week as an adult, assuming it evades predation, makes its dawdling even more dangerous. Mad in pursuit of light, Lunas lose time from their task and can even injure themselves, beating their wings against a light source. Ferreting the Luna out of the shadows, light can also make it more vulnerable, especially to the attention of its most faithful admirer, the bat.

I turn the security light off, once June arrives, leaving only the front door light on, my antidote to the darkness that descends on the mountain, dense with the unknown, full of strange rustling sounds, and even those that I can identify—the yipping of coyotes—seem too close. There is still a risk in this, at least for the Lunas. Opening the front door in early morning, I have found—more often than I would like—two pale green tails lying on the deck, as well as the shattered remnants of green wings. Sometimes that is how I know the Lunas have arrived, by finding one sacrificed in this way. I assume it alights on the screen door and, transfixed, is an easy target for an enterprising bat, which devours the Luna's luscious body. Its tails, which the bat rejects, are key to the Luna's in-flight defense, a crafty piece of its adaptive puzzle.

A long-held belief is that the Luna's long, streaming tails divert attention away from its body (much as eyespots do, though, in the case of the bat, which is not primarily a visual hunter, these would not deter it). Biologist Jesse R. Barber and his colleagues at Boise State University have conducted experiments explaining exactly how this works. The Luna's tails are in fact "auditory deflectors" that confuse the bat's sonar: "The spinning hindwing tails of luna moths lure echolocating bat attacks to these nonessential appendages in over half of bat-moth interactions." In essence, the tails provide "an anti-bat strategy designed to divert bat attacks," Barber posits. Though I am disquieted when I find the remnants of my favorite moth, the Luna, these days I am also comforted, even reassured. It means I have a bat—at least one—still living on the mountain.

When I first came to the Catskills in 2002, bats were among the most vibrant constituents of the ecosystem. I would see them at dusk as the weather warmed, pursuing insects overhead, their flight jagged, skittering. My evenings were filled with bats. If I ventured outside at night, they would swoop down, flying low enough for me to hear the rasp of their leathery wings. New York is home to at least nine bat species—six, including the most common, are cave bats, which overwinter here; the other three, tree bats, are migratory. Of the cave bats, the most populous and familiar is the little brown, whose body length is a mere two inches, with a wingspan of eight to nine inches. It was this bat that popped up behind a shutter as I was power-washing the house one warm afternoon. First a hook-like "thumb" appeared, then another, followed by a dark-brown gremlin's face featuring funnel-shaped ears, a pointed snout, and tiny, sharp teeth. The bat looked at me, blinked, a tuft of soapsuds topping its fuzzy head. I had awakened this critter, washing it out of its afternoon roost. The bat flew off into the woods, leaving a haven that had been carefully chosen for its western exposure, facing the afternoon sun, to keep it warm.

I would, as penance, erect a "bat box" on a nearby tree, painted black to retain heat and facing west, as a summer roost for the little browns, which winter in caves and mines where they hibernate, but seek out the crevices of trees, rocks, and houses, too, in summer. It was soon tenanted by a small colony of about ten bats and I would be spellbound, watching them fly out of it at dusk to hunt, then return at dawn to sleep away the day. Bats are warm-blooded, the only mammals to truly fly, featuring a thumb and four fingers on each wing (with a bone structure recalling the human hand), which they use to cling to trees and walls. Still, there is something so primordial about them, as much myth and legend as reality, living shadows of the night.

But the bat house has been empty for years now. Starting in 2006, the number of bats in the Catskills began to decline. A mysterious disease, known as White-nose Syndrome (WNS), which has been identified as a fungus, revealing itself as a white residue on the muzzle and wings of afflicted bats, has devastated hibernating bat populations. Nearly six million bats have perished since WNS was discovered in an upstate New York cave, perhaps the result of a foreign organism tracked in by cavers. But it is the little browns, my tenants and neighbors, that have suffered the most, losing, by some estimates, 80 percent of their population. This epidemic, affecting numerous cave bat species, according to the Center for Biological Diversity, is "the worst wildlife disease outbreak in North

American history," and despite ongoing efforts to find a definitive cause and cure, "shows no sign of slowing down."

Bats are crucial insectivores, easily the most overlooked—for their economic value to humans—wild critters. It is estimated that one little brown bat can eat up to 600 mosquitoes in an hour. This is a boon to humans, especially in an era of deadly mosquito-borne illnesses, including the West Nile virus and eastern equine encephalitis. Bats also act as a critical control by gobbling up insects that eat or damage crops. Researchers estimate "the value of pest control services provided by bats in the U.S. alone range from a low of $3.7 billion to a high of $53 billion a year." Their loss, if WNS continues to spread, will doubtless have severe and lasting impacts on ecological systems as well as agriculture. Bats also, as insect eaters, save the environment further degradation by lessening the application of pesticides for exactly this purpose.

Extinction, as a concept, seems theoretical and remote until you realize, with a jolt, that the critters around you are disappearing. The little brown bat is on the verge of extinction in the Catskills and the entire Northeast. The loss of the bats would be like the loss of the night itself.

THE FEEDERS! I HAD FORGOTTEN TO BRING THEM IN. THE THISTLE seed feeders were the only ones I hung in summer, for the goldfinches, which regurgitate this seed to feed their young. Most other critters, even birds, eschew thistle seed as distasteful. But there was one black bear, a young female, that had taken a liking to it, and having lost too many feeders to her bear brethren, I decided to fetch them, even though it was well after midnight on the mountain.

The back porch light was bright, lighting my way. I hadn't found bits of Lunas left by the bats by this light, which was why I had started leaving it on. But heading up the deck steps, I was startled to see a Luna moth caught in a spider web, woven by a clever arachnid to trap insects attracted to the light. I put down the feeders, seeing that the moth was still alive. Delicately, I removed it from the web. Holding the Luna in my hand, I saw strands of the web were stuck to its wings, which I had to remove so that the moth could fly. Drawing a deep breath, anxious that I might tear the delicate wings, I painstakingly peeled off the remnants of the sticky web. The operation was a success and I opened my fingers so that it could fly. But the Luna stayed in my hand, and I could see every intricate marking so clearly, including the purplish border that edged its forewings, a stunning complement to its iridescent green.

The moth seemed content in the glow of the porch light, which spilled over the steps where I was sitting. Perhaps, too, it was comforted by the warmth of my hand. I knew I should shoo it away, go back inside, turn off the light, and let the Luna do its urgent work this warm June evening. But for just a few minutes more, I wanted to sit there with the moth, to feel the lightness of the Luna's being, its weight no more than a breath, to hold beauty in my hand.

CHAPTER FIVE

The Quarry Fox

(For Eric Ashby, the man who loved foxes)

O N THE DESK WHERE I WORK SITS A SHARD OF BLUESTONE. IT is slight but strong, sharpened to a point. I found it when I was clearing the creek, embedded in a bank. Its shape is suggestive of an arrowhead and I kept it for that—a tangible link to the Lenni Lenape, whose forebears were the first to work stone in these hills. It has become a talisman of sorts; when I hold this small grayish stone, obdurate, a survivor, it is as if I am touching the hardscrabble past of this place, the Catskills, which is so bound up with bluestone.

When I first explored this land, I found jumbles of rock in the nearby woods—the foundations of forgotten structures, once part of the family farm that endured here. A sap house, I suspected, had graced one such pile, set among sugar maples, where I also discovered curious circles of stone, whose use I could not reckon. Later I would learn that these modest medallions might be pot lids, thin pieces of bluestone, cut to cover earthenware jars, perhaps, in this case, containing maple syrup. But when I returned in spring after a winter away to claim them, they had disappeared, lost to nature's sleight of hand. Such humble items, pot lids, but they tell the tale of how people lived, how they used their resources. Pot lids—and even the arrowhead I prize—offer a connection with those who came before, a sense of continuity, history evinced in the ordinary of everyday.

Nowhere on Catskill land is that history—the saga of settler and rock—more manifest than in the bluestone fence. There is an art to its construction, a mastery in the way the rocks are assembled, without mortar to bind them, and a pride in their placement. Even if disheveled after so many years, some losing their moorings, these walls of rock retain their integrity—as boundary markers and even today, to pen in farm animals. Cows, mostly well behaved, especially the dairy breeds, usually can be contained by stone walls, three feet tall or more, with a flat top layer of rock (though a rambunctious young bull will jump a fence in search of sweet, untried clover and his freedom). But to keep in the sure-footed, stolid sheep, farmers would finish an even higher fence with stone set vertically, to prevent those capable climbers from scaling the walls—goats, nimble, ever curious, and willful, could not be deterred from breaking out even by bluestone, but they were relatively rare then in Catskill farm country. The fences also served as useful repositories for rock that had been cleared to cultivate fields or to dig out foundations, some of which would find its way into farmhouse floors and as lining for cellar walls. (They also defined the intimate boundaries of family graveyards, cool, shady places, some of which survive today, often visible from the road, the past in our present.) Seen from a distance, stone walls crisscross hillsides so prettily, and few sights are as pleasing on a Catskills summer day as a herd of black-and-white Holsteins, each with its jigsaw puzzle of patches, ambling down the mountain, in green, unbroken pastureland, framed by sharp diagonals of bluestone fencing.

Sometimes when I am walking in the woods, I come across stone walls, or the remnants of them, now adorned with wild rose, host to wildflowers, edging the perimeter of fallow fields. Chipmunks hiding in the rocks herald my arrival, their shrill chirps issuing a warning to others. One pops out of the wall, eyes me, then retreats with scattershot quickness. "You know," I tell them for the umpteenth time, "if you didn't announce yourself, I wouldn't even know you were here."

They are guarding their home, this fortress of rock, from any number of predators—the sharp-shinned hawk, low-flying and agile, perched in a dead oak I passed on the way, the red fox I glimpsed, staying still in the underbrush, patiently waiting for some small rodent to venture forth. Mice live here as well as chipmunks, visiting voles, a toad perhaps, escaping the afternoon sun, tiny shrews, and all sorts of insects. Arachnids, too, abound. A daddy longlegs, commonly known as "harvestman," with its tiny brown button of a body and its eight long lissome legs—more companionable than spiders, its close relatives, as it doesn't bite us—has hitched a ride on my shoulder, probably when I sat down on the wall to rest, which I won't realize until I return home.

Snakes are also aficionados of rock, where they seek prey as well as sanctuary from predators. In the Western Catskills, I am most likely to stumble on garter snakes, bold enough to hiss but harmless. (One sleek green garter snake, trimmed with three horizontal yellow stripes, spent a summer curled in a bucket of bluestone scraps, content to monitor the mouse population.) Timber rattlesnakes especially are fond of sunning themselves on stone walls, cold-blooded creatures leaching warmth from rock. The venomous snakes, with their signature spade-shaped head and coy warning rattle, seem to have a particular affinity for bluestone.

Timber rattlers still tenet the slopes of Overlook Mountain near Woodstock in the Catskills High Peaks, where the commercial bluestone industry started up in the mid-1800s. Excavations at the eastern front, in Ulster County, yielded the handsome bluish-gray sandstone prized by architects such as Stanford White, who incorporated it in his plans for many notable buildings, including the legendary Louis Comfort Tiffany house in Manhattan, the mammoth Romanesque mansion completed in 1885, a lost icon of the Gilded Age. Its strength and durability, as well as its beauty, turned bluestone into sidewalks and curbs in many American cities until Portland cement, cheaper and more accessible, effectively ended the industry at the beginning of the twentieth century. The quarries of the Eastern Catskills are now still, quiet, eerie places, with clean cuts of stone lying where they were left, as if work ceased yesterday, though they have been idle for decades, now part of the Forest Preserve of the Catskill Park, protected from further exploitation. Today, hikers ascending the mountain need to be as chary of timber rattlers as the quarrymen of old, whose watchword was "If y' hear a rattler, ya got bluestone." Though in decline in the Catskills, timber rattlesnakes still patrol some abandoned quarries, now ceding to nature, colonized by birch and butternut trees, rooting in the rocky soil, and they have taken up residence in the ruins of the old Overlook Mountain House, as confident as sacred serpents, basking in the pathways that approach its crumbling buildings.

When I rest my hand on "Old Blue"—my favorite snag of rock, light gray, the color of sandstone in the Western Catskills, flush with hues of dusty rose—I can feel its strength but cannot fathom its ancientness. Old Blue was born over 300 million years ago as sand deposited in a shallow sea that covered much of this region, runoff from other mountains—the Acadians, whose heights once touched 28,000 feet, which vanished millennia ago, leaving behind the building materials for bluestone, the new enduring generation of rock. Tracing what looks like ripples with my finger, I consider that these marks on Old Blue's surface could be fossilized water—the ghostly imprint of that prehistoric sea that turned sand into bedrock eons ago.

The boulders that erupt on Catskill land all tell this story. But some have their own stories, too. Most bluestone is composed of horizontal layers, easy to separate and ready to work, which makes it such an attractive building material. But Old Blue's laminations are vertical, giving it the appearance of a prodigious loaf of sliced bread. Time has gentled it, crowning Old Blue with wildflowers that thrive in its thin, scattered soil—daisy fleabane, butter-and-eggs, devil's paintbrush with its flaming orange face—now a sunlit roost for mourning doves, an occasional oriole adding color, its "slices" of stone beginning to fall away, the rock, cracked and corroded, soothed by dark green moss, stippled with pale lichens, ancient organisms even older than the rocks. Reading Rock, crouched in Little Porcupine Creek, is named for the seeming natural seat set in the boulder, making it an appealing place to sit and read. But the seat is cut out so expertly, its straight back strict, that I wonder if this wasn't the work of some long-ago farmer, intent on a perfect square of bluestone to use as a step—perhaps for his wife to ascend to a wagon or a child to stand on while drawing water from a well.

Of the many boulders on this land—Turtleback, Scylla and Charybdis, Compass Stone, to name a few—perhaps the most ordinary, at least at a glance, is Woodchuck Rock. It is a low, flat slab in gradations of gray, darker where the rock is exposed, with splotches of muddy orange, hardly a handsome stone. What gives Woodchuck Rock its story is that it has a cleft in the middle, separating it into two large stones, fairly equal in size. Nature, ever adaptive, saw an opportunity in this. It had the makings, the woodchuck decided, of a perfect winter den. Where the stone splits open, there is a gap over a foot wide. (The rock, despite its short stature—only one and a half feet tall—is still a respectable size, measuring eight feet long, and more than five feet wide.) The first fall I lived on the mountain, walking over to check on the winter feeders, which are in the woods just beyond, I noticed a pile of fresh dirt, newly dug, by the rock. The woodchuck, I realized, was excavating, and had made that gap in the rock the entrance to its den, which would be several feet below the surface, and extend more than fifteen feet underground.

I was amazed at the mound of earth and broken rock accumulating at the entrance to the burrow, which I would visit when the woodchuck, now growing ever fatter, gorging for its long winter sleep, would scamper off to forage. It is a comic critter, the woodchuck, standing like a sentinel to survey its surroundings, as is characteristic of squirrel family members, including its western cousin the prairie dog and the clan's smallest member, the chipmunk, resident of bluestone walls. Its roly-poly body and short legs have earned the woodchuck—a large rodent with grizzled grayish-brown fur and a stumpy dark tail—its most common name "groundhog," immortalized by the day that every year celebrates this critter's putative talents as a meteorologist, February 2nd. Still paying tribute to its porcine appearance is the nickname "whistle pig," which, looks aside, refers to the woodchuck's penchant for whistling a high-pitched warning when threatened. The woodchuck, despite inspiring the popular ditty "How much wood would a woodchuck chuck, if a woodchuck could chuck wood?" does not "chuck" wood, that naturalists know of, at least. The name "woodchuck," which I learned as a child, probably derives from *wojak*, the Lenni Lenape name for the animal, revered as a tribal ancestor.

What distinguishes the woodchuck, other than its many names, is its powerful front feet, equipped with long, curved claws that are well adapted for digging, and its strong rodent's teeth that can bite through the toughest roots and dig out the most recalcitrant rocks. Before the snows came, the woodchuck was denned in, and if I worried about a predator gaining access to its winter retreat, there was no need. The woodchuck had pushed up dirt from within the den, so that the opening was sealed shut. The deep Catskills snows would soon make the burrow even more impenetrable. Woodchuck Rock was as fortified as King Agamemnon's Palace at Mycenae, guarded by its own Lion Gate.

After several seasons the woodchuck left or perhaps it was taken by coyote, bobcat or the swift red fox. This was, after all, only its winter den. Woodchucks have different summer homes—this one had a particular fondness for the drainage pipes that channel water away from the dirt driveway, disappearing in one whenever I appeared or at the slightest sound of any activity. (Animals on the mountain seem wilder than their compatriots in the valley. Friends living there report woodchucks that don't flee—and crows that don't fly away—at the sight of humans, perhaps more resigned to the presence of people if not more trusting of them.) I would worry, in a heavy downpour, that it might

drown or be washed away into a gulley, but the woodchuck would always reappear, soaked, its spiky fur giving it a decidedly punk appearance, to graze on clover flowers, purple and white, whose admiration it
shares with butterflies and honeybees.

That spring after the first winter without the woodchuck sleeping
in its den, I was startled to see, in early April, a red fox lurking around
Woodchuck Rock. It was a splendid fox, with its rusty red coat, high
black boots, and brush of a tail tipped with white, but still, my heart
sank, thinking perhaps I'd been wrong—that the woodchuck was still
there and was now being stalked by the fox. My fears seemed confirmed
when I saw the fox scratching at the den's entrance and then wriggling
its lithe body into the opening between the two half boulders. I watched,
through binoculars, holding my breath. Then the fox emerged, without
the woodchuck dangling in its jaws (a large prey for the slender, fine-
boned fox, but too succulent to resist), and retreated into the woods,
which thicken a few yards from Woodchuck Rock. I thought no more
of the fox and its failed hunting expedition until several days later, when
I noticed the fox, now coming and going, back and forth, to the rock.
The sight of a fox was not unusual. What was strange, or certainly new,
was that I was seeing it regularly during the day. That can be concerning,
as foxes, especially in areas where humans are—who often persecute

them as pests, perceiving them as predators on livestock, especially poultry—have learned to hide, which is consistent with their shy nature.
They are also "crepuscular," meaning most active at twilight, between
the sheltering hours of dawn and dusk, though they will also hunt their
primary prey, small mammals, by day. A fox in the afternoon—or any
change in its usually reclusive behavior—gives me pause, because they
are also a principal rabies vector. But this fox was showing no signs of
erratic behavior, and seemed purposeful in its task.

It occurred to me, finally, that this could be a fox family that had
taken over the woodchuck's carefully constructed den. Critters are opportunistic recyclers—bluebirds will tenant old woodpecker holes (if
they are not squeezed out by starlings, known to surveil woodpeckers at
work, then steal the cavity once it is completed), and gray squirrels will
reuse them too, for their winter home—and with some redesigning that
usually consists of widening the den, a woodchuck burrow can make a
cozy fox nursery. Red foxes, I knew from the musty smell of the courting
male that fills the winter woods (which in summer I would take for a
skunk, offering its opprobrium), mate in January and February, their
thick, luxurious fur allowing for love in the snow and cold so that their
kits are born in spring, when prey to feed their hungry young—mice,
eastern cottontails and chipmunks—are plentiful. The gestation period
for foxes, approximately fifty-three days, would bring their birth to just
about now, the middle of April. I wondered if the fox I was seeing was
the male (known as "dog fox" or "Reynard") bringing food to his mate
as she nursed their young, probably four to six pups. Father Fox, as fox
fathers famously are, was a devoted spouse and parent.

It was, in a way, an odd place for them to den. Woodchuck Rock
is in sight of the house, not far from the dirt driveway. They had to be
aware of me. But I am a practiced quiet presence on the mountain and
in spring especially, when birds are nesting and babies are being born, I
try to be unobtrusive. I couldn't believe my luck, actually—I was looking
forward to chronicling the kits' adventures, to watching them grow. And

Woodchuck Rock did have its advantages. It is just down the slope from the pond, giving the foxes a source of fresh water (and a breakfast of green frog, if they were quick). The woods are nearby too, which offered cover—a place to flee, if necessary—as well as the "edge habitat," that zone of grasses, shrubs and small trees between forest and field, so conducive to wildlife, where cottontail rabbit, the fox's favorite fare, and other small mammals, abound. Deeper, but not far, into the woods is also the stone wall, still a boundary marker, built so many years ago, which would be hiding other rodents for the foxes to feed their brood until they were old enough for both parents to teach them to hunt. So perhaps the foxes, renowned for their cleverness, were even smarter than I knew. I marveled at Woodchuck Rock, old bone of this mountain, at how it protected life—the woodchuck in its winter slumber, now sheltering the vixen and her kits. I could not wait to see them.

Midway through May, I was coming down from the pond—not far from Woodchuck Rock, which had been quiet, without the presence of either parent—where I'd headed up to assail a tangle of wild roses, my quixotic quest. At the entrance to the den, I was surprised to see two pudgy fuzzy pups. They would be about a month old, still wearing their brownish-gray pelage, starting to play outside of the den, taking in the world. I stood still, watching them. I saw one pounce, perhaps on an insect—a cricket perhaps, the kit's first prey. I was enthralled. Then the mother appeared—her soft orange-ish fur the color of Creamsicle, glinting in the morning sunlight—at the edge of the woods. She was carrying what looked like a red squirrel in her jaws, breakfast for her young. Seeing me, she withdrew into the thick underbrush, probably still watching but no longer visible. At her quick retreat, one of the kits, closest to the entrance, perhaps not as bold as the other, disappeared into the den. Then the remaining kit, to my surprise, started to run toward me, taking short, bumbling steps. It looked like an exuberant puppy, happily greeting its new owner. I was alarmed—I did not want this little fox to be so sure of people. Suddenly it stopped. I remember looking right at it, our eyes met. I saw a shift there—that openness, that innocence, even, was re-placed by something else. A realization, perhaps an intuition that we call instinct, that this new creature standing before it was not a friend. I'll never forget how that little critter's expression changed, a sudden aware-ness of danger. For the first time in its life, the fox felt fear and I was the cause of it.

The kit bounded back to Woodchuck Rock and found the safety of its den. I hurried into the house to allow the vixen to return to her brood. How would this episode affect her? By the next morning, perhaps even that afternoon, the foxes were gone. I watched for either mate for several days but there was no sign. The mother had moved her family to another den. Red foxes have several, as havens to escape danger—to escape a predator, such as I. It was a bittersweet ending to the story of Woodchuck Rock, which has had no tenants, not even an indolent wood-chuck, sleeping away winter, since. A leafy green shrub I cannot identify is growing thick and high in the mighty cleft, still a host to life, and I wonder if I should clear it, and the earth collecting there, in case another critter wants to set up residence. But most of all I wonder about the look in that young fox's eyes. How did she or he, with no experience of hu-mans, know to fear me? When did that kick in, that notion, and why? What changed that seemingly careless little kit into a wary one? I was relieved that the fox had turned away, for I knew it was best for the an-imal not to be so trusting. Even so, I felt a certain sadness, and a lingering sense of loss.

THERE ARE MANY QUARRY ROADS ON THE MOUNTAIN, THE CATSKILL legacy of mining. Some are so overgrown, they seem to go nowhere. Oth-ers lead to small haphazard excavations, marked by scattered cuts of stone. Still others, rutted tracks, form a network that connects this em-pire of forgotten quarries, paths once taken by wagons, later by trucks, to move blocks of bluestone. It took me a while to ferret them out, using my straight, strong walking stick "Trusty Tree," a maple sapling deer had stripped of its bark, to part the tall weeds as I explored, my anchor in the loose, rocky soil.

A steep switchback, zigzagging up the mountainside, takes me to Lazy Hawk Ridge, where there is an old quarry about an acre in size. I enjoy visiting it. Abandoned quarries are, for me, places of solace. In a way, they remind me of archaeological digs, not only for the disruption of rock, the displacement of earth, but for the silence that lingers over them—these sites that have seen so much human activity—and for how quickly nature reclaims them. Since I first saw the quarry over ten years ago, tender young saplings have toughened into trees—hardy red oaks and supple yellow birches, their leaves shimmering in the light summer breeze. A stand of staghorn sumac, weedy scavenger of waste places, a shrub that can grow as tall as a tree, is encroaching. In fall, the undistinguished sumac's leaves will turn the color of flame. Wildflowers, ever scrappy, have a home here—Queen Anne's lace, its flower of the purest white, as delicate as a doily, so refined in these rough surroundings. Spikes of mullein are everywhere, some stalks as high as six feet, each bursting with rows of tiny yellow blooms. A medicinal plant, beloved of herbalists, mullein heals landscapes, too.

The quarry itself is eight feet deep, gouged out of the mountain, its layers of bluestone exposed. Standing at the edge, I can see broken brown glass—beer bottles probably—littering the quarry floor, shards of a timeless rite. As I descend the dirt ramp to the bottom—built to haul out rock—I find, where the weeds rise up, earth meeting stone, an assortment of appliances, some half buried, as large as a refrigerator, as small as a fan; a car bumper, rusted out; a blue plastic wading pool filled, somewhat satirically, with soggy "Posted: No Trespassing" signs; a metal milk can, souvenir of another era; even a baby carriage, relics of future digs. Today I discover, as I sometimes do, underwear—a bra and a pair of boxer shorts, draped, insouciant, over a tire. Such stories, these silent stones could tell . . .

What alerts me isn't a sound. It is the sense I often get in the woods of being watched. I look up. A red fox is standing on the dirt ramp, no more than six feet away. My first feeling is fear and I grip Trusty Tree with both hands. I have no idea how long it has been watching me, what it wants. What concerns me is that it seems to have sought me out, has come so close. Foxes usually flee as do bobcats, coyotes, too, if I stumble

upon them. (It may take a shout to roust them, but even black bears will gallop off.) But the fox is showing no signs of aggression. I could chase it away but I don't, reassured by its calm demeanor. Its coat, the color of embers, glinting in the afternoon sun, its legs, as black as char . . . I wonder briefly if animals know how beautiful they are, if they are aware of their infinite grace.

The fox does not avert its gaze, its eyes meeting mine, but I see no challenge in them. It is curious, as I am, but "curiosity" seems too slight a word, too animal, to describe what I see in its amber eyes. The look is searching, almost as if the fox is trying to understand . . . Abruptly it turns, lopes up the ramp with that light, loose gait, reminiscent of a cat's. At the end of the ramp, it pauses, gives me a backward glance, then is gone, as silent as a shade.

DANGER
Keep Out

That sign, deep in the woods, marks the beginning of a new kingdom, the realm of the mine, the working quarry that exists on the other side of the mountain. When I first came, it was a modest mine, about four acres, in sync with its surroundings, the bluestone it offered a gift. Since then, the quarry has expanded—it now claims over ten acres, creating a high, broad rent, stripped of trees, visible from the valley, the rock laid bare.

Seen from above—from "Old Road," running along Lazy Hawk Ridge, which once ferried bluestone out of the abandoned quarry—the Big Quarry bears some resemblance to a vast ancient amphitheater, its ledges of rock, from where I stand, looking terraced, reminiscent of rows of seats cut into stone. Work in the quarry has fallen off of late, activity has lessened—vagaries of the economy, the rise and fall of bluestone— leaving a bulldozer astride a pile of rubble, a steam shovel with its jaws clasping rock, an excavation still life.

Every spring I clear paths through the woods of fallen limbs, branches shattered by winter and shifting rocks that emerge from the mud. Critters, as we do, take the easiest course, and I often find the tracks of coyote and fox, an occasional bobcat, and turkeys, too, the stamp of their feet everywhere, tokens of their appreciation. One such path leads to "The Pike," as it is called, widened in recent years to accommodate trucks hauling slabs of bluestone out of the Big Quarry. As work has slowed, the volume of traffic decreasing, weeds are already overtaking the dirt road and rain has filled some of the deeper ruts, creating shallow pools where I am startled to see the faces of tiny green frogs peering out at me as I pass.

Two crows call back and forth as I hike The Pike on my way to the mine, which has its own fascination for me, this city in the woods that is also a ruin. The crows, among other topics, are probably announcing my arrival in their mysterious language of caws, gurgles and rattles. There is no human activity here that goes unnoticed by many eyes, where critters are always watching for predators, and for prey. I am still disquieted by my encounter with the fox several days ago—not by its proximity, for I realized the fox wasn't rabid. But by what I saw in its eyes—those amber eyes glittering with intelligence—that longing, that desire to comprehend, perhaps even to connect. Had I seen in the fox's eyes anything other than what I wanted to see? Can we ever read the expressions of wild animals fairly, without attributing to them our own emotions? I once encountered at the pond a big, brawny coyote—clearly showing the interbreeding with wolves that has made eastern coyotes so much larger than their stringy western counterparts—which held its ground as I approached, running only after I shouted at it, waving my walking stick. It looked at me, I thought at the time, with a defiance that edged on a challenge, such sullenness in its stare. But, were these qualities even there? Or is that just how I see this animal, which I begrudge for its boldness, for how close it comes to the house, even during the day, and for its taking of so many small mammals, especially spring fawns?

I hear a loud thrumming—the sound of a generator powering the stonecutters' saws. The Pike takes me past the sign warning of danger, tacked to a tree, to where the Big Quarry begins. It is marked by the suddenness of sky, trees along the road and higher up having been felled, lost to the bulldozer, the crush of rock. Looking up, as I follow The Pike along the quarry's periphery, I can see a high hill of tailings—the rocky remnants of quarrying. The tailings form a ridge of refuse—shale, its broken red pieces scattered everywhere, sandstone fragments, of varying shapes and sizes, some large blocks, cracked, discarded for their imperfections, mixed with imposing mounds of excavated earth. Even though bluestone, laid in horizontal slabs, can be separated with relative ease, it is usually wedged between layers of soft shale and mudstone—a mixture of clay and fine-grained silt—sedimentary rocks devoid of commercial value. The proportion of "waste" to "useful product" is ten to one. This mine, when it eventually closes, will be "remediated"—restored, to some degree, though of course not to its pristine state. The mountain, this side of it, has been forever changed.

What must it have been like, when so many Catskill hills were marred by such mines, seen from afar, as this one is today, as a deep furrowing scar on the mountain, and marked, too, by clouds of dust, the hazy residue of saws slicing through rock. In the Eastern Catskills, where the bluestone industry thrived in the nineteenth century, ramshackle settlements sometimes grew up next to the piles of tailings, housing the Irish immigrants who came to work the mines—work that could cost them limbs, sometimes their lives, the price of cutting stone out of the mountain, trimming it with hammer and chisel, loading it onto wagons to be hauled away.

I am always uneasy, walking in the shadow of the sloping waste rock. Rain, a summer torrent, perhaps even the cascade of thunder, could jostle dirt and stone, causing a shift in the unstable rubble, sending it tumbling down to the road below. I keep my eye on it for safety's sake, aware of bluestone's danger, the price of its beauty, mindful of such a slide. Suddenly I grow tired of this landscape, impatient for the perfection of the forest, where I rarely feel such apprehension, even as day starts to recede and I hear the rustling of animals ready for their evening rounds. Gazing up as I begin to head back, I see atop the pyre of tailings, standing so lightly not a stone is sent skittering, a red fox, watching me.

Even on the coldest days, the crows call. They are the voice as well as the eyes and ears of the mountain. They telegraph my presence with harsh staccato bursts when I throw out bread for them in summer or leave unsalted peanuts in the shell under the feeders in winter. They are mindful of predators, especially in nesting season, and their cawing always brings me to the window, usually to witness a red-tailed hawk being rudely escorted out of the crows' territory—which comprises Lazy Hawk Ridge where they have their roost—not only with raucous insults but even an occasional airborne assault. Just recently, I saw two crows chasing away a larger black bird, noisily berating it. A raven, I realized. Ravens have made significant inroads into the Western Catskills in the past twenty years, since the turn of the new century. Still, I had never seen one here, assuming they preferred the higher elevations of neighboring areas. But observing this scene, I wondered if the murder of crows that command this Catskill foothill have banned its larger corvid cousin from taking up residence here.

Several days after I saw the fox at the Big Quarry, I was awoken at dawn by a loud, hoarse scolding, the mantra of the vigilante crows.

The cawing was so close, right outside of my bedroom, which overlooks the wildflower garden, a small mound that marks where the holding tank for my supply of spring water has been set into the earth. I have let wildflowers grow up there, not only to disguise the tank but as a refuge, a mini-ecosystem that shelters various small critters: chipmunks (which scurry back and forth to the shed several yards away, where they hide in the labyrinth of bricks that serve as its foundation), white-footed mice, voles and also birds, as showy as indigo buntings, as modest as chipping sparrows, perched on bending shafts of goldenrod and milkweed to feast on an endless array of insects. That morning, in the dim light, as I peered through the window, which serves as a sort of blind, I was startled to see the fox. It was sitting at the edge of the wildflower garden with a posture similar to a cat's—erect, its bushy tail wrapped around its front paws— ignoring the crow's protestations. They were perched in the nearby blue spruce, as well as on the roof of the shed—I counted twelve in all— venting their disapproval.

The fox was so still as it sat there, seemingly nonplussed by the crows' attentions. I suspected that less than indifference or equanimity, what the fox was showing was patience. It was simply waiting for something to emerge from the wildflower garden, which, I assumed, it had already discovered was a repository of prey, a veritable snack bar of small mammals. What also struck me—what always surprises me about the fox—was how small it was. An adult red fox may average eight to twelve pounds and vary in total length from forty-eight to fifty-seven inches, its long fluffy tail accounting for nearly half of that, according to the New York State Department of Environmental Conservation. Standing about fifteen inches high (at the shoulder), it is slightly smaller than a medium-sized domestic dog, which, as a member of the Family Canidae, the fox is related to. The red fox's persona seems so much larger than that, perhaps because of its flame-colored coat, so bold, so *visible,* which really makes no sense in nature. What can be the adaptive advantage in that? Other carnivores that share its habitat—the coyote and bobcat—

are grayish brown and mottled tan respectively, colors that allow them to blend in easily, in virtually any habitat, day or night. Even the red fox's cousin, the gray fox, with its grizzled grayish coat, a ridge of black running down its back and ruddy underparts, seems better adapted both to hunt and to avoid being hunted than the American red fox, a subspecies of *Vulpes vulpes*, which is regarded as the most successful wild carnivore worldwide, having colonized much of North America, Europe, Asia, Northern Africa, and Australia (where it was introduced for hunting). And yet, even with its striking red coat, the fur so thick, so soft, as this fox's was, which has made it prized for its pelt as well as a ready target as it roams open fields in search of the same critters it was waiting for now, the red fox has managed to endure on this earth for thirty million years.

It was June, still nesting season, but the red fox—as opposed to the gray fox, with its long, hooked claws—cannot climb trees, as its distant cousin can, for safety from coyotes that prey on both or to steal eggs and young from nests. The crows, regardless, were right to identify the red fox as a predator, among the most skillful hunters in the natural world, though perhaps this grudge was personal. The crows, arguably the most intelligent of birds (along with their fellow corvids, the ravens), have perhaps met their match in the fox, whose craftiness is celebrated in one of Aesop's most popular fables, "The Fox and the Crow." In it, the fox tricks the crow out of a piece of cheese by praising her glossy feathers, her bright eye, and, most fatuously, her singing voice, which the fox tells her he longs to hear. The crow, unable to resist the fox's blandishments (how human is that?) starts to caw, dropping the cheese, which the fox snatches and runs off with—the moral of this particular story: Do not trust flatterers. What intrigues, from a naturalist's perspective, about this tale, is that "Aesop" (or whoever contrived these fables, supposedly in the 7th century BCE) pinpointed the quality that is most associated with the fox—its cleverness. To praise, of all things, the crow's hoarse, coarse singing voice is not only funny but illustrates the vulnerability—of all of us—to flattery, what a seduction that is. The fox itself is also susceptible to vanity, in that other Aesop favorite, "The Fox and the Grapes." A fox, unable to snag a bunch of succulent grapes, ripening high overhead on a vine, after several failed leaps, stalks off saying "Those grapes are probably sour anyhow"—the moral being, we often devalue what we can't have. Here, Aesop shows himself to be an astute observer, noting the fox's fondness for fruit and its ability to jump (as high as fifteen feet), which is a vital aspect of its hunting prowess. Intelligence and vanity, qualities of the fox in these fables, are also eminently human, as is its adaptability. The red fox, as are we, is an omnivore, which is key to its success in surviving in so many environments, from country to seashore to suburban backyard.

Despite the crows' squawking, the fox stayed still for some time, transfixed by the wavering tall grasses of the wildflower garden, alert to any opportunity. I was struck by the particular vividness of this fox's coat, its rich red—the scientific name of the eastern red fox, *Vulpes vulpes fulva* (*vulpes* equals "fox," and *fulva* derives from *fulvus*, meaning "reddish brown")—and "red" foxes can also have color morphs, ranging from a yellowish or golden hue to silver or even black. But this fox's coat seemed especially lustrous, a searing red, and I wondered if this fellow sitting so stoically—its grayish white chin, throat and belly exposed—could be the same fox I had sighted at the quarry. There are always foxes on the mountain. But coyotes are the dominant carnivore here and they are the fox's enemy—hunting it as prey, and thus eliminating a competitor for the eastern cottontails and other small mammals that the fox also prizes. Where there are coyotes, the fox tends to stay away, reluctant to breed in an area where its young, especially, would be so vulnerable to predation. Perhaps that was why I had seen it in the vicinity of the quarry (and why that fox family had denned so near the house)—the human presence and attendant activity would help keep the coyotes away, which travel in a pack, whereas the fox, mostly solitary except in the mating season, would find it easier to hide, and survive, so near people.

The crows had kept up their tirade for a full ten minutes, the fox stubbornly sitting there, until finally, perhaps figuring that this ancient nemesis had alerted the mice to its presence, it abruptly stood up and turned to go. The crows, their chorus triumphant, perhaps even mocking, followed it as it started to saunter off. The fox did not run, almost as if it didn't want to give the crows any satisfaction, even pausing to lick a paw (reminding me of my cats, trying to act nonchalant), which drew one bold crow even closer, swiping at its head with its beak. The fox snapped at it, almost catching the bird in its jaws—had that casual pose been a ruse, perhaps, of the trickster fox? Now the crows, indignant before, became irate, and the fox began to run with the crows in pursuit, its light, sleek body seeming to float in the air as it bounded away, finally reaching the safety of the woods. "That old wildflower garden probably had nothing good to eat anyhow," I could almost hear the fox thinking.

THE FOX LIVES, IN A SENSE, ON THE EDGE—ON THE FOREST EDGE where it prefers to hunt, seeking the critters that inhabit this transitional zone between woods and field, and on the edge between night and day. "Crepuscular" almost seems too muscular a word to describe such a delicate time, when afternoon dissolves into night, and morning begins to scatter darkness. Dawn is at first almost imperceptible, a subtle shift from black to dour gray, in summer in the Catskills, often mixing with a swirling mist, luminous in the growing light, until the sun, finally ascending the mountain, disperses it. Dusk is perhaps truer to crepuscular's origins—deriving from the Latin *creper,* meaning "dark, gloomy"—when light is lost, seeping away, finally resolving into a soft, irresolute glow, the sun's remains, its refracted rays as it slips below the horizon, a time that once served as a warning to humans—soon it would be night, when the world was no longer theirs.

It is primal, the fear we have of the forest at night—a fear that is perhaps only deepened by moonlight, which casts shadows that flicker and dart through the leaves, dappling trees that, in the dimness, appear grotesque, contorted, as large rocks seem to resemble sleeping creatures, about to awaken. A walk in the nighttime woods, even with a strong light in hand—a must to traverse the roots and ruts that in the darkness are even more treacherous—does little to assuage this unease. Shining the light in the direction of a sudden noise—the snapping of twigs, in the daytime barely noticeable but now seeming to ricochet around me—I am met by two greenish eyes glowing in the dark. I freeze. I have ventured into the woods on a whim—not far, I am too much of a coward. Still, I am afraid. What this critter is, I cannot be sure. A coyote, out hunting, would probably be with its pack—this animal, I think, is alone. A bobcat, always skittish, would have already fled. Deer are not so delicate in their movements—large critters, with hooves, I often hear them crashing their way through the woods. Standing there, I have to suppress something more than fear—a visceral sense of vulnerability, that awful awareness of just how inadequate we are, as humans, on our own, in nature. I want to run. But I know if I do, I will fall flat on my face on the rocky ground. And if this animal is a bear (which I doubt, because the eyes seem too low—a bear, even on all fours, would be taller), running could trigger its prey response. An aggressive black bear would pursue and overwhelm me easily.

I shift the light, catch a flash of red. A fox, out patrolling the woods from dusk to dawn, searching for prey. I am relieved. Suddenly the glowing eyes are gone. But I have heard nothing, not even the crackle of last year's leaves, still littering the forest floor. I swing the light in a wider arc but there is no sign. The fox has melted away, disappearing into the darkness. The only sound is the wind, rustling the leaves, the creaking of branches, then, overhead, as I hurry home, the eerie whinnying of a screech owl.

It was dusk the next time I saw the fox. I often sit on the back steps in summer, watching the twilight deepen as day turns to night, escorted by the lingering evensong of the wood thrush and the twinkling of fireflies. I have thought, seeing the lightning bugs, as I called them as a child, shimmering in the high meadow, pranks played on the darkness, that fireflies may be the origin of what people, in rural cultures, called fairies or pixies, lights dancing in the shadowy woods. They are magical, perhaps even more for me now that as an adult, as my curiosity and urge to catch them have given way to wonder, even though I know that this is their courting ritual, how fireflies signal and attract a mate. It is such a gift of nature, this luminous silent spectacle, heralding summer as the silvery chorus of the peepers, those tiny tree frogs, announces spring.

This is also the fox's time, a creature in between day and night, forest and field, appearing, now disappearing in the murky light as night descends, making us wonder if it was there at all, a trick of the eye or perhaps even a spirit, as some have believed, this shape-shifter of the woods. So I wasn't surprised when I glimpsed the fox again, this time emerging at the opening into the woods, just as the fireflies were beginning to rise. It saw me first—at least it was looking at me when I spotted it. I stayed very still, and the fox, perhaps reassured by the distance between us—at least a hundred feet or so—started to trot with the light, languid gait so reminiscent of a cat's along the edge of the upper meadow. At the pond it paused for a drink, then disappeared from view, perhaps scouting out the frogs or even the red-winged blackbird nests, though I heard no scolding by those feisty, combative birds.

That sighting was the first of many that summer—the red fox often appeared at dusk as I sat on the back steps. It would look at me, almost as if acknowledging my presence, and increasingly comfortable, would take its time journeying to the pond, which seemed to be its destination. I had the opportunity to see the fox hunt and I was stunned. It looked, I thought, like a cat—even my own two felines, those stalkers of feathers and hapless moths—crouched, tail flicking, slinking low to the ground, then pouncing, leaping into the air and coming down on its prey, holding it with its paws. I would watch the fox at work, fascinated, through binoculars that allowed me to observe its hunting rituals. I even observed, on several occasions, the fox seeming to play with its captured prey, as cats are wont to do. My first thought (fanciful but I couldn't resist) was that the fox—with its movements and even mannerisms I had observed, as elegant, as fine as a feline's—was some sort of hybrid. The fox is classified as a canid—its large triangular ears, long, pointed snout, and forty-two adult teeth (adult cats have thirty) mark it as a member of the Family *Canidae*. Dogs also have non-retractable claws. But true to its in-between nature, the red fox has the cat's ability to partially retract its front claws and its feet are soft and hair-covered, adaptations that aid in the fox's stalking.

But it is the fox's eyes that truly distinguish it. The pupils are vertically slit, like a cat's, which allows the eye to sharpen its focus, especially useful in low light when the fox hunts, enabling it to distinguish shapes in the darkness, small critters skittering across the forest floor. The fox also has a "well-developed" tapetum lucidum, a layer of connective tissue "on the innermost sheath of the eyeball" that reflects light back out of the eye, intensifying the images the fox sees, enabling it to see in low light levels, a trait it also shares with cats. Wildlife biologist J. David Henry, whose fieldwork studying red foxes in Saskatchewan's Prince Albert's National Park has afforded him an unparalleled intimacy with and knowledge of these critters, though pointing out that the red fox is a "bona fide member of the Family *Canidae*" and noting the extensive commonalities the fox has with other canids, both physical and behavioral, also observes these additional similarities to cats: long feline whiskers (longer in foxes than in other canids) on both muzzle and wrists, which serve as "tactile organs"; long, thin, strong teeth that resemble the cat's "daggerlike" canines, which the fox uses to kill prey by biting, not shaking as dogs mostly do; and a "lateral threat display," conjuring images of the scary Halloween cat, with its arched back, fur erect, standing on end, and stiff-legged fearsome stance. The fox and small cats, such as the bobcat and my own two wild cats, those so-called domestic shorthairs, hunt the same prey. Their similarities in behavior and appearance can be attributed to "convergent evolution," Henry concludes, where two different species look and act in similar ways because they occupy the same ecological niche.

One behavior that the red fox shares with its canid cousins is caching its food. The fox, with its small stomach, needs to eat often and will often bury its kills, saving some for later, as a way of assuring it will have enough to eat. I may have stumbled on one such cache on one of my hikes to the Big Quarry. That summer, I would often sense that the fox was with me as I walked through the woods. I rarely saw it, if ever, except as a fleeting movement out of the corner of my eye. If I turned to look, there would be nothing. But I felt the fox's presence, and we had gotten fairly easy with each other. From its first visit—that I was aware of—at the wildflower garden in June and then several days later, when it would appear at dusk and lope along the edge of the upper meadow, searching for mice and other small prey before it disappeared at the pond, I had become accustomed to the fox and clearly, it had gotten used to me. I felt privileged to see it up close and to observe it, almost acting as it would on its own. On my hikes, especially to the quarry, I had that sense of being watched, and I fancied that the fox was hiking along with me.

Once, before I reached The Pike, on my shortcut through the woods in September when the leaves had already started to fall in the Catskills, I came across—I actually walked into—something hard and substantial hidden under a pile of red and yellow sugar maple leaves.

Uncovering it, I found a coyote skull, partially covered with dirt, bone white, with a tuft of grayish hair still attached to the top. It was magnificent, a flawless specimen, and I happily put it aside on a low rock as a way of assuring I would find it again on my way back from the quarry. It would make a handsome, even prized, addition to my collection of feathers (turkey, crow, blue jay and one slender gray catbird feather); several bird nests that the wind had blown down from trees after nesting season, including a tiny hummingbird's, the size of a teacup; a paper wasp nest I had evicted from the mailbox, fortunately abandoned; and several butterflies, including a tiger swallowtail I had found fallen in the grass, as perfect in death as it had been in fluttering life.

But on my return, which was no more than an hour later, the skull was gone, vanished. I checked all the rocks in the area, just to make sure, but found nothing. I knew no one—no one human, that is—had passed this way. It was a trail I shared with turkeys and deer and an occasional coyote, judging from the tracks I would find in the mud, which could last into late summer, the ground is so wet with springs. Then it occurred to me that perhaps I had found the fox's cache, that it had covered this rather large (certainly for it) treasure with fallen leaves and dirt, to be taken away later. The fox had merely reclaimed it. Or was this Reynard the Fox, the trickster of fable, having a joke on me? I had to laugh. But more than anything, I took this as confirmation that the fox was there, my instinct was right—the fox was indeed the companion of my days, as well as of dusk.

OCTOBER IS THE MOST HEARTBREAKING MONTH IN THE CATSKILLS— even more than April, with its promise of spring revoked by a sudden late snowstorm. As early as August some maples start to turn, especially those that are distressed if it has been dry and hot. It always tears at me, that first intimation of fall, which will soon settle over these hills, though I am comforted by the prospect of seeing the trees, stalwart green through summer, ablaze with color—the red, yellow and orange of the sugar maples, the red maples' scarlet, the golds of the aspens and birches, the beeches and oaks, amber and bronze, subdued but strong, all anchored on the hillsides by pine and spruce, hemlocks too, which, ever green, reassure us that life will endure. There is a moment, in October, when these colors converge, even though many leaves have already fallen, the rich deep musk of decay rising from the forest floor, especially heady on a warm, languid Indian summer day as leaves mix with the earth, the fragrance of the autumnal woods. That moment, lasting no more than a few days, with luck, when these colors, so vibrant and seemingly immortal, can be lost overnight—to rain and wind or a sudden cold snap. That is the heartbreak, waking up to leaves wrenched from trees, which are suddenly exposed, stripped, cadavers awaiting winter.

It was then, in October, that I visited the old quarry, to see how it had fared over the summer, if my favorite "archaeological dig" had accrued any new artifacts, to see it once more before winter, reassured, as I always am, by how nature is overtaking it. Also, the forecast called for rain, perhaps even snow showers, a storm predicted to bring wind from the north, which would take down much of the foliage, clinging tenuously, even now, to the trees. Snow here can come as early as October, usually around Halloween, and though rarely deep it is a warning worth heeding, that winter will have its way, rendering the mountain impassable. My stay at the quarry was brief, after checking to see what new "souvenirs" had been left and noting, with satisfaction, there seemed to be none (it is "my" land, though I have given up any hope of policing it).

As I started on Old Road to pick up the switchback where I would head back down the mountain, I saw the fox. It was standing in the middle of the road, looking at me with those amber eyes. I paused, then started to walk, expecting the fox to run. It took a few gamboling steps, then turned back, again looking at me. We proceeded like this on Old Road for ten minutes or so along Lazy Hawk Ridge, where that side of the mountain slopes down into the Big Quarry. I got the sense the fox wanted me to follow it and I did. Finally it turned off, to the left, after casting me one last glance. I turned off as well and found myself standing among tall weeds, waist high. But when I looked more closely, I saw that beneath the grasses and low shrubs was the outline of a path, rutted with wheel marks—yet another quarry road, one I didn't know. I kept walking on it, intrigued, but the fox had disappeared.

Then I entered an area overcome with tall pines—the ground was littered with needles, making it such a soft place to step, and the air was redolent with the pungent scent of resin. Bluestone rocks of varying sizes were scattered everywhere, on both sides of the path, as I walked under the deep green canopy, beside aspen and birch trees, their leaves a warm golden hue. Finally I came across a small bluestone quarry. It was tucked into the side of the mountain, a pocket quarry, so discreet it was, and diminutive, sheltered by pine and spruce trees, deep with newly fallen leaves, still flush with color. There was no remnant of any human activity, not a single item. No one, it seemed, knew about this place, not anymore. From this vantage I could see, lower down the mountain, through the trees, my own house, which looked startlingly small from this height. But I had never seen the "secret quarry"—as I would call it, and where I would return often—from below. It was pristine, this place, protected from the elements—the evergreens offered a break from the wind, protection from rain and snow as well as any human eyes—a repository only of rock, tumbled pieces of bluestone that had been left behind, home to the crows and the fox.

Had the fox really led me here? Did it want me to follow it? Or had I bumbled onto this place, an accident, which had just happened to be in the fox's path as it went about its business? I had to ask these questions as a naturalist—I had to try to avoid sentimentality and projection, attributing to the fox my own subjective interpretation of its actions. But I knew in my heart that, that summer, the fox had shared its world with me and as winter approached—and with it, the return to our separate worlds—the quarry fox had given me this one last gift.

THAT WAS THE LAST TIME I SAW THE FOX THAT FALL. THE EVENINGS, darker ever earlier and chilly, kept me inside at dusk, and an occasional glance out the window did not reveal the fox at twilight. The first snow fell just two weeks later, then there was a reprieve through November, until the snow began to fly in earnest in December. I waited, as I always do, until Christmas to put out my feeders, convinced that by then the black bears—the males especially, who den up later than the females, pregnant, or with first-year cubs—would be tucked into their dens. That winter descended suddenly, with wind and cold that freezes your face and splits your fingertips, and the ground did not show, finally free of snow, until March. On my winter hikes or on deeper days, snowshoeing excursions, I looked for the fox's dainty tracks but never found any. Which wasn't really a surprise. Its light, furry feet, narrow and small, do not leave as clearly defined marks as other critters, especially in deep snow. I often found the coyotes' trail but there were usually several traveling together, the pack disturbing the snow. But that winter, in January, I did smell a fox—its musky scent, akin to a skunk that has sprayed—marking his territory, searching for a mate. And, at about that time, I heard that plaintive, yowl-like cry that pierces the winter darkness, haunting, unearthly even, a lone fox calling out in the night.

It was early May when I began my vigil again, sitting out on the back steps in the crisp, thin twilight, listening to the peepers celebrate spring with their whistling calls, waiting for the fox to appear at the Cut. When I didn't see it after several days, I began to worry—though why, I wasn't sure. Perhaps the fox was busy with a new family. I hadn't really seen it last year until June, so despite my unease, I told myself that there wasn't any significance in this, other than I missed the fox, my friend, my companion of the woods. I hadn't been off the mountain for nearly two weeks, watching, as I was, all the activity, the business of birds especially—tree swallows and bluebirds, busy at their nesting boxes, robins and phoebes building nests beneath both the back and front deck, chipping sparrows coming and going in the Alberta spruce, which seemed to house a song sparrow family too, and the hummingbirds, whose territorial squabbles had already prompted me to put out two more sugar water feeders in a failed attempt to thwart their fighting.

At the bottom of the road leading down the mountain where I pause before turning left or right, across from the pasture where Farmer McCagg's herd of handsome Hereford cattle roam, I saw it—a small reddish brown bundle of fur lying by the roadside: a fox, hit by a passing car, forever stilled. Was it my fox? I didn't know—my fox or not, it was such a sad sight, such a waste of life, of beauty. I sat in the car for some time, fighting back tears. I just knew, suddenly just knew, that the fox—my fox—was gone.

Later that day, I hiked through the woods, clearing the path of winter as I went, hoping against hope that the fox would find me. But I had no sense of it, I did not feel its eyes on me, though I kept looking, turning, hoping to find its amber eyes. As I turned into The Pike to head to the Big Quarry, I heard a truck rattling up behind me. I paused by the side of the dirt road, waiting for the old red pick-up to pass. It was after five, when most work was done for the day. But this was Sean, the foreman, a burly, good-natured guy, from a long line of Catskill bluestone quarrymen. We often talked when he would find me hiking and compared notes on the fox, which was well known to the quarrymen, whom Sean described as "so friendly, you half expect him to raise a paw and wave at you when you drive by."

"I guess the fox got hit," I began when he stopped. "I saw it down the mountain, across the road."

"That wasn't the quarry fox," Sean said, and my heart leaped. "The quarry fox got shot."

"What?" I said, incredulous.

"Local kid . . ." Sean went on, sitting behind the wheel of his truck. He was looking straight ahead, as if he didn't want to meet my eyes. "Shoulda known better, meant well, started feeding the fox. It wandered into the wrong backyard, in broad daylight, just sat there. Got too used to people, I guess. This guy came out of his house—has chickens, kids—thought the fox was rabid, shot him dead." He shook his head. "Can't blame him, really. Big mistake, feeding that fox."

I just stood there, too stunned to speak.

"You know what I hate," Sean said, turning to me, and now I could see that his eyes were glistening. "I hate it when animals get hit in spring, after all that cold and snow, surviving that, making it through winter. Then right off the bat, first thing, bang, they get hit, killed, crossing the road . . . It just seems so damn unfair." He paused. "That's what I hate, more than anything."

Things I have learned living in the Catskills.

I have learned that nature will overcome.

I have learned about the power of water.

I have learned the wind is ceaseless.

I have learned that cold can burn.

I have learned that death is common; it is life that is extraordinary.

I have learned to always wear denim and boots, a wide-brimmed hat and leather gloves when working—and hiking—on the mountain.

I have learned it takes true courage to love wild critters.

CHAPTER SIX

Autumn's Leave-takings

(For Roger Tory Peterson)

THE CHANGING LIGHT—THE EVER-SHORTENING DAYS—IS THE signal for the summer birds to begin their migrations south. Temperature alone is too variable and, as a result, deceptive, occasionally luring younger, untried birds into staying too long. A mild October can quickly turn blustery, with railing winds that thwart flight, and a sudden frost will kill insects, a primary food, especially of seasonal residents. The nectar-loving ruby-throated hummingbird, which is also an insectivore, is the first to heed the warning of the light's diminishment. By the end of August, it has departed, this smallest of birds, weighing less than an ounce, embarking on a perilous journey that can take it to southern Mexico and Central America, some crossing the Gulf of Mexico in a single flight. The males, with their ruby throat, are the first to leave, followed several days later by females and adolescents, all an iridescent green, their feathers shimmering in the sunlight.

The feeders I first put out for the hummingbirds in May—sugar water to sustain them until the insects rise and wildflowers bloom—I still hang in summer and into September, in case any are migrating through, searching for sustenance. But the truth is, I like the hummingbirds' company. Seeing them at the feeders, so close to the house, I get to know them as individuals—their quirky and passionate personalities —follow their battles for territory and mates, witness their endless squabbling, and, if I am lucky, hear their soft squeaks of delight. When the hummingbirds leave, they take with them summer and even though September is perhaps the most beautiful month in the Catskills, with skies that are startlingly blue and uncharacteristically cloudless, the loss of the hummingbirds, so quicksilver, glittering with life, hints at winter.

SOMETIMES IN SUMMER IN THE CATSKILLS, THE MEADOWS HIGH, fragrant with sweet grasses, vivid with wildflowers, it is difficult for me to imagine winter, and easy to deny that it will come. How, I wonder, standing at the meadow's edge, can this lush landscape become barren, then stilled in white when the snow flies, the ground frozen into March. Above all, it is the busyness of birds that I cannot imagine this world without—the tree swallows diving for insects over the meadows, the phoebe's incessant calling, *"fee-bee, fee-bee,"* the robin strutting on the lawn, cocking its head to ogle grubs lurking in the grass. Long before winter sets in, stealing away summer's warmth and lingering days, they will be gone. But nature, ever ironic, perhaps, provides a brilliant recompense for autumn's leave-takings. The lessening of daylight and September's chilly nights quiet the chlorophyll that gives leaves their green color, key to the process of photosynthesis, allowing the vibrant colors of fall to emerge. In the Catskills, the maples lead the way—the red maples in scarlet, some leaves a palette of colors, yellow and orange-red painted on green, and the eclectic sugar maples, their foliage a raucous collage of red, orange and gold, set the mountains and valleys afire. These colors,

in their defiance, seem to me a surge of life that must take the trees'
last strength as they prepare to enter dormancy, their winter sleep.
When the leaves finally fall, taken down by wind and rain in October,
they leave the trees as bare as corpses, returning to the earth to build
a new layer of soil.

On this chilly early October morning, the light so bright,
edged with the clarity that autumn brings, I first heard, then saw a bluebird perched on the flagpole. His soft, low *"tu-a-wee"* call alerted me.
Then, through binoculars, I spied him, the royal blue of his back and tail
feathers marking him as male. Was it this summer's Poppa Bluebird? Or
perhaps one of the five youngsters that fledged in June? The bluebird pair
had returned a couple of weeks after their first nest emptied, looking to
start their next brood. In past years, the parents have made their second
nest of summer in this same cedar wood box, the one that faces east, its
opening turned away from the hot afternoon sun. Momma had carefully
examined the box, peeking in and even disappearing inside, and I expected
them to start anew. But perhaps because the tree swallows, which they tolerated the first time, were still buzzing around their own box, ten feet or
so from the bluebirds' digs, they demurred. And the swallows did harry
them, swooping down on the bluebirds, then landing on their former
neighbors' home and peering in—how rude is that? Much to my disap-
pointment, the bluebird pair flew off and I didn't have the pleasure—and
the worry, to be frank—of watching over them during a second nesting.

   When the bluebirds fly off of the flagpole, they usually dive
down to the lawn, which abuts the lower meadow, to pluck a grub or
some other insect from the grass. If pickings are slim, they will fly
away on another hunting expedition, seeking food to feed their demanding young. This time, though, Poppa Bluebird flew straight up.
Up, up, on a vertical line, and he kept climbing. I tried to follow him
with my eye, then grabbed the binoculars to catch sight of him, as he
was quickly fading from view. But even with the binoculars, I could

only see a momentary speck in the deep blue October sky, and then he disappeared.

I had the sense that Poppa Bluebird had left—that he had just begun his fall migration (and none too soon, as the previous night's temperature had fallen to 25 degrees, the first hard frost of the season, glazing the meadows, and my hardy mums, with a sheen of icy white). I felt so privileged, honored, really, to see his departure. As I stood there, silently wishing him luck, knowing the many dangers he would encounter on his journey south, and then hopefully, on his return trip next spring, I wondered if he would be the lucky bluebird who would make it "home" to breed here, in this box that I tend so carefully. And I took comfort in the fact that in this numbers game of species survival, the box that "my" bluebirds have now favored for three years, through several generations, which I meticulously prepare for them every spring and guard protectively through the summer, has helped the bluebirds to fledge over twenty nestlings.

P OPPA BLUEBIRD'S LEAVING DID REMIND ME TO CHECK THE FOUR nesting boxes—to clean out whatever nests were still left, and to ensure that no other unwelcome tenants had taken up residence there. That way, next spring, I wouldn't have to fret over when the new nesting birds would arrive. This past year, the bluebirds surprised me by arriving earlier than before, in March instead of April, and I was glad that I had readied their box for them the previous fall.

The bluebirds' own box, the one that faces east, along the edge of the front meadow, which I monitor so carefully, I had already emptied, and it was clear of any other inhabitants and relatively clean (if one doesn't count the white feces stains smeared on its sides, which actually are an indication that the bluebird young have successfully fledged). The box opposite it, about ten feet away, which the bluebirds' feisty tree swallow neighbors favor, still held its messy nest, adorned with all sorts of feathers, the swallows' trademark. (I cleared out that old nest and walked with it some distance, where I threw it into the underbrush. Even though there were no longer nesting birds, I did not want to attract any predators to the boxes next season.)

I also found, attached to the underside of the roof (which is why I wear leather gloves to check), a paper wasp nest, relatively small, with several wasps still in it. I pulled out the delicate grayish nest, often described as resembling an "upside down umbrella," and threw it into the meadow (the wasps, already sluggish, will die before winter). Fortunately, paper wasps, which like to tenant nesting boxes, are relatively mild in demeanor (they will usually sting only if threatened). Paper wasps, oddly, at least to us, are attracted by the odor of propane. I often find several dead wasps in my gas stoves every winter, where they sneaked in seeking shelter—and in summer, they routinely make their papery nest on the underside of the metal cylinder that covers the gauge for my thousand-gallon fuel tank—the four-legged "pig," as it is affectionately called in the country, for its porcine appearance. I admire paper wasps—they are elegant insects, with their slender, dark-brown body cinched at the waist, speckled with bright yellow spots, as well as clever architects and master builders. If I didn't knock down their pretty nests, they would adorn every eve of my house in summer. And though always wary of them, I also appreciate paper wasps as pollinators whose grubs are eager eaters of caterpillar pests that can strip deciduous trees of their leaves.

Then I walked out to the upper meadow and opened a third box, which faces west, to find the remnants of yet another swallow nest, festooned with feathers too, and when I reached up inside it, I also found a wasp nest, this one abandoned. Just one more box to check . . . but when I opened it, at the edge of the woods, facing east, I was stunned. There within were two little tree swallow nestlings, stiff and desiccated. They were long dead, as the swallows depart the mountain to flock in the valley by mid-July, after all their young have fledged. What, I wondered, had happened here? Had both, or even one, of the parents died? And there were only these two babies left, no other eggs were in the box—usually tree swallows lay as many as four to seven eggs, pure white, with their characteristic pointy end. Had the other young swallows fledged and these nestlings been abandoned? I thought back to spring, couldn't remember any exceptional cold snaps after the swallows' arrival (they tend to come in early April with the bluebirds) or long chilly, rainy periods that might have limited insect prey for the parents to feed them.

The tree swallows, songbirds too, are such graceful, exuberant flyers—I love to watch them pursuing insects in the air so extravagantly, with their acrobatic twists and turns, racing each other, twittering loudly, often breaking off and flying at the house, recklessly so, at the last second veering up and away, showing off, entertaining themselves (and certainly me), then diving dramatically, and starting their hunt—or is it a game?—all over again. I did remember seeing a pair of tree swallows perched on that nesting box in spring, having claimed it, even going in and out, I assumed, to tend to eggs, and then their young.

I took the baby birds out of the box, still in their nest, and I left them in the nearby woods. I knew that any number of critters would find them and, never fussy about food, would eat them—that would at least give some meaning to these lost lives, by making them part of the cycle of life itself, these two little tree swallows, who had never known that joy of flying that is their birthright and their brilliance, whose discovery had filled me with such sadness.

Bluebirds, redux: Several days later, on a warm afternoon, an Indian Summer day, with temperatures in the seventies after autumn's first hard freeze, I looked out on the lower meadow and, to my amazement, saw a small "flock" of five bluebirds. One sat atop the flagpole, two others were perched on each nesting box, another was sitting in the small, struggling Japanese maple tree that I still need to surround with burlap before winter sets in to protect it from the ravages of hungry deer (they will strip its sweet bark as well as its tender twigs), and still another I saw in the barberry bush, happily eating its red fall berries. I counted three males, all of whom were paler than the bluebird poppa I had bid adieu to, and two even paler females. Were these the five baby bluebirds I had seen fledge this spring, zooming out of the box in the front meadow on their first flight, now all grown up, finally about to take their first fall journey south?

Or were they the offspring of that second nest that I did not supervise this summer (probably to the bluebirds' relief), after the parents—specifically the female—chose a different new home, probably in a hollow of a dead tree in the woods? I could hear their soft calling and although they did not stay for more than a few minutes, they lifted my spirits after my discovery of the sad little tree swallow nestlings. Especially the sight of the two females busily inspecting the bluebird boxes (one female hopped into one, while the other poked her body so far into the other that only her pale blue tail feathers showed), unable to resist checking out these spiffy properties as a prospective place to raise a family, their first, next spring—reminding me of my mother, who had to visit every model home she ever encountered, a memory that made me laugh out loud.

I felt such a sense of pride, seeing them, "my" spring brood, as I now claimed them, convinced they had returned before heading south.

Safe journey, my bluebirds, and hurry home, hurry home . . .

THIS MORNING, A FLOCK OF ROBINS FLEW IN, THEY WERE EVERYWHERE, all around the house, males, females, and juveniles, banded together for the migration south. Or at least south of here, the Catskills, since robins only fly so far south as they need to find food. I have never seen robins this far north in winter, but they do tend to migrate later than other birds, readily switching their diet to fruit from worms and grubs, though the weather has been mild enough for them to still poke around in the ground for worms, their favorite protein. It rained overnight and the robins were drinking eagerly from the puddles of fresh, cool water. Later in the day, after the robins left, a troop of dark-eyed juncos arrived, the grayish sparrow with the white outer tail feathers that flash as it flies, a crisp, cool little bird that also breeds here. But the flock that arrived today, I figured, had headed in from the north, to overwinter in more temperate forests than those of Maine and Canada. It was a joy to see them, especially after so many summer birds have traveled south, feasting on the seeds offered by the dried grasses still standing in the meadow, reminding me that life will persist here, that even in the coldest months, this mountain is home to many.

BEFORE SEPTEMBER WANES, BEFORE THE HUMMINGBIRDS START THEIR journey south, autumn's first leave-taking, as early as July, is birdsong. It is the only time I dread on the mountain, when these voices still. And though it is still summer, the meadows high with wildflowers, wholesome and fresh, a greeting every day, my mood is not lightened. It isn't winter that the cessation of birdsong presages, with August still to come, its hot, sunny days conjuring the clatter of an occasional cicada, as well as autumn's burnished afternoons, crisp, with the scent of apples in the air, and languid Indian summer days, thick with light. I feel a loneliness that not even winter, its silence absolute, save for the desolate wind, instills. When the birds stop singing, I lose their companionship, an intimacy with their lives, a ready familiarity with their everyday that I treasure.

But birds sing for their own purpose, indifferent to my attachments. They sing to establish territories and to win mates—their brightly colored plumage also announces their eligibility to females and warns away other males. The red-winged blackbird, first to arrive in March, flashes his scarlet epaulets when challenged by a rival as well as while courting, whose triumphant "*konk-a-ree,*" as he alights in the cattails, proclaims that the pond is his.

More than 270 species of birds have been sighted in the Catskills region, most of them summer residents, nesting in the mountains, in an astonishing array: from the feisty ruby-throated hummingbird to the fearsome red-tailed hawk, patrolling the mountain for prey; familiar birds such as the American robin and eccentrics like the American woodcock; warblers, so many types, like flickering lights in the woods, their songs sweet bits of sound; and the great-horned owl, its haunting call—"*hoo-hoo hoooooo hoo-hoo*"—resonant and deep, filling the nighttime forest. There are flashy birds, true to their name—the rose-breasted grosbeak, indigo bunting, and scarlet tanager—and modest ones too, though the song sparrow's "stickpin" marking its breast adds a jaunty dash to its brownish coat, and its bright, cheery call, so welcome after winter, is an icon of early spring; noisy birds, the pileated woodpecker, largest in North America, with its flaming red crest, whose drumming reverberates through the forest, advertising his territory and appealing for a mate; brilliant, beloved bluebirds and the aerial artists, tree swallows and barn swallows too, flashing in the morning light; purple finches, the "sparrow dipped in raspberry juice," and goldfinches, those convivial "flying daffodils"; gray catbirds with their black cap and russet rump, mewing from the wild rose, cousin to that other mimic, the brown thrasher, singing from its repertoire of over a thousand songs while mocking other birds; and the shy wood thrush, my favorite songster, its melancholy melody played at dusk, a flute echoing in the woods, regretful of the end of day.

By mid-May or so, most have arrived, joining the red-winged blackbird, spring's solitary sentinel perched in the aspen tree, guarding the three nests his females have woven into the reeds. The rakish redwing is polygynous, meaning he has more than one mate and often many. But females are also enterprising—DNA testing has shown that not all the eggs in these nests may be his. The mandate of every bird—every critter—is species survival. And just as the male, ever vainglorious, is eager to assert his genes, females, desirous of improving their offspring's pedigree, are willingly seduced by an ardent songster and feathers that are carefully coiffed.

Many birds do pair-bond for a season or even beyond—if they escape predation and the perils of migration (though flirting seems to be forgiven). Poppa Bluebird offers his mate tasty morsels as she sits on the nest, the two of them then working together to feed their ravenous nestlings. But the rapscallion ruby-throat breaks his bond immediately after mating, abandoning the hummingbird female to build a nest and raise the young on her own—then will quarrel with his "ex" and his fledglings at the feeder, chasing them away, territoriality trumping family ties. Red-tailed hawks, which "mate for life" (or at least until a partner dies, when the survivor seeks another spouse), are perhaps more pragmatic than romantic. Having the same mate can save time, especially in constructing their mammoth nest of dried sticks, which the pair may refurbish and reuse the following spring. And as egg incubation is fairly long (averaging thirty-five days, as opposed to twelve to fourteen days for the bluebirds), and youngsters only leave the nest after about six weeks, the earlier the red-tails breed, the better.

The nesting season offers us a glimpse into the private lives of birds, to which we are alerted, in spring, by birdsong. On May and June mornings, I wake very early, just as night is dissolving into gray. I lie in bed and listen, waiting. The robin is often the first to sing—"*cheerily cheer up cheer up cheerily*"—one, then several, vying to start the new day. The red-winged blackbirds join in, many with nests in the meadow grass, their harsh, nasal trilling in counterpoint to the robins' lyric song. As dawn seeps over the mountain, the violet sky turning pinkish, rosy toes touching the morning, everyone is singing, each song telling a story, offering an invitation, issuing a warning, all proclaiming, *I am here!*

Their singing ebbs as their activity lessens, mostly after the nestlings have fledged. But birds do not fall completely silent. What ceases for most birds is their vibrant courtship song, their proclamation of territory. They will still use a variety of calls to communicate, notifying each other of any threat, food, or their location, especially when flying. Venturing up the hill to the pond in spring, I am confronted by the red-winged blackbird, the male and his mates, scolding me with "*chack!chack!chack!chack!chack!*," rapid and intense, signifying danger. (Big Red will dive-bomb me too and

any other intruder, several of them joining together to "mob" potential predators—crows especially, hawks and blue jays, and even harrying hapless deer who stray too close.) And I can also hear the red-wings commenting to each other throughout the summer—a sharp "*chek*"—as they forage for food. As summer progresses, birds need most of their energy for the molt, when old and damaged feathers are replaced by sturdy new ones, in preparation for the rigors of migration, and for those that stay in the Catskills, the harsh winter to come.

One bird, though, resumes his singing, just as the others are ending theirs. The shrill blue jay is silent in spring, guarding the secret of his own nest while stealthily raiding those of other species, stealing their eggs and young, to feed his offspring. Nature, in its irony, signals the end of nesting season, its sweet musicality, with the caustic calling—"*Jaay! Jaay! Jaay!*"—of that wily trickster, the blue jay.

It is already mid-October, and with the exception of a few stragglers, probably adolescents slow to start their first migration, most of the summer birds have left. But autumn, such a glittering, golden season in the Catskills, has bestowed one last gift . . .

A great blue heron has been visiting the pond for nearly a week, in the morning and again at dusk. I was awestruck, the first time I saw it there, standing in the pond's shallows, stock still, as is characteristic of how herons hunt. That pose that it strikes always reminds me of representations I have seen of how Native Americans, when our world was theirs, hunted for fish in the pristine streams and rivers of this continent, so still and quiet, spear drawn, ready for whatever fish, finally accustomed to their presence, would come too close. Here in the Catskill foothills, along the banks of the Delaware River, whose branches join in Hancock, New York and then flow as one into Pennsylvania, the Lenni Lenape would have been those Native Americans who fished following the heron's stately and lethal example.

When I first saw the heron, walking up the dirt driveway early in the day, I was so excited at the sight of this imposing bird—which can be over four feet tall, the tallest and largest of North American waders, its legs reminiscent of a stork's, with its long, thick bill, as sharp and as dangerous as a dagger—I ran to get a better look at it. First lesson: don't run at a great blue heron—it took off immediately, startled by my charge. I was afraid that I had frightened it away. But then, that evening, I saw it again at the pond, this time from my back deck.

I stood, as still and as silent as the bird, just watching it seemingly do nothing . . . The heron was standing at the pond's edge, where the water, at this time of year, ranges from a few inches at the bank to a shallow foot or so, before it plunges into a depth of about ten feet in the pond's center. In fall, pond levels are lower than they would be in spring with the snow melt and warm rains. But it still falls off quickly—spring fed as it is and dug deep enough to both shelter overwintering largemouth bass as well as to serve as a "fire pond"—vital, as there are no hydrants on the mountain, in case of an emergency. I am grateful for the latter. The former, though, the pond as a residence of largemouth bass, irks me continually. The previous steward of this land could not bear the calling of green frogs, which are very populous in the Catskills, and had the pond stocked with largemouth bass to counter them. (Green frogs, often mistaken for bull frogs, are similar in appearance but smaller, about two and a half to three and a half inches compared to the bull frog's imposing three and a half to eight inches—from the tip of the nose to their amphibian anus—and their call is a plunking sound—"*gunk-gunk-gunk*"—while the bull frog croons its deep, mellow "*rum, rum, jug o' rum.*")

The largemouth bass, a popular game fish here, are also voracious—they will eat anything, insects and small birds if they can catch them, even each other, and, as a result, the green frog population in my pond—eggs to tadpoles—is greatly diminished each year. To aid the frogs on land so they would be less vulnerable to hawks and mammal foes such as foxes and raccoons, I let a border grow up around the pond's perimeter. Grasses, small shrubs, and wildflowers (including delicate blue violets in spring; then through summer, daisies as well as their smaller cousin, common fleabane; Black-eyed Susan; red Sweet William; foxglove; wild geranium, in white and lavender blossoms; yellow wood sorrel; and in fall, my favorite, New York aster, with its violet-blue flowers) provide the frogs a place not only to hide from predators but also to seek shelter—with their sensitive, moisture-loving skin—from the bright sun. Still, though, the frogs are vulnerable, as I saw when, approaching the pond one morning, I was greeted by a small black reptilian head popping up like a periscope through the grasses at the pond's edge—a Northern water snake, which hunts both in the pond and on land and is itself vulnerable to predators such as raccoons. At the sight of me, yet another predator, it slithered into the pond, leaving behind barely a ripple.

The calling of the green frogs—they start their *gunk-gunk-gunking* as if they are plucking the bass string of a banjo, in June, just as the chorus of spring peepers begins to fade—is a refrain that comforts me and lulls me to sleep as sure as the falling of soft summer rain. Still, female green frogs can lay a gelatinous blob of 1,000 to 4,000 eggs at a time and then also another before the breeding season is done—and so many, if they successfully morphed into frogs from the tadpoles that they will become, would soon overwhelm most ponds if they were not plucked by predators and people too, who often glean at least some egg clusters every spring. But in the case of my pond, which is small by Catskill standards, about twenty feet wide by thirty feet long, when I do find those gobs of eggs (usually attached to cattail husks, algae filaments, sticks or some other scummy detritus), I am thrilled. The largemouth bass do make quick work of most of them, but enough tadpoles survive to adulthood to at least call from the banks during summer, to breed the next generation, and under cover of the wildflowers, to sit in the shade, jumping into the pond with a loud "plop" and an occasional disconcerting cry that has given them the nickname "screaming frogs" whenever anyone comes too close.

The bass are the most insolent creatures—they eyeball me defiantly when I spot them in the water. Are they thinking of assailing me, too, as I have seen them waylay dragonflies when they fly too close to the pond's surface, jumping out of the water to snatch them from the air? They certainly seem brazen enough, these assassins. Friends have suggested that I fish them out—but I don't have the heart for that in spite of myself and once, when I did manage to capture one with a net, I couldn't kill it. (Instead, I left it in a bucket of pond water, convinced that by the time I returned, some critter, a black bear or certainly a raccoon, would have absconded with it, which I regarded as a victory. Instead, several days later, I found the fish still alive, still looking up at me impudently. Defeated, totally, I dumped it back into the pond.)

So my hope, of course—and part of why I was so elated—at the arrival of the great blue heron was that this accomplished fisher would soon dispatch my largemouth bass. Not all of them, of course. But some of them. (Maybe just one? Or even two, with luck? Please . . .) The great blue hunts a variety of fish, reptiles, amphibians, and even other birds and small mammals, stabbing them with its bill, then swallowing them whole. Though it can choke on larger prey, I trusted that this great blue (which is actually more blue-gray in coloration) could devour my own particular brand of bass, which I suspected were eating each other in the confines of my smallish pond before they could get too big—the largest of which I had ever observed was ten inches or so, small by large-mouth bass standards, which can grow over twenty inches long. The heron will stand in the shallows, waiting for fish to come too close— finally trusting that its long legs are reeds—within its range, which is rather copious, considering the length of the great blue's sinewy S-shaped neck and that flashing bayonet that passes as its bill. When the great blue strikes, it is with a sudden deadly precision that our Lenni Lenape hunter admired so and no doubt emulated, spear poised for the kill, watching perhaps for a largemouth bass, which is at least a species native to North America.

Stoical and aloof, the way it stands at the pond's edge, silently waiting in the shallows, everything about the great blue heron is, to me, majestic. But for those that it hunts, its slow, swift movements, its every regal, still step, are only calculated to bring death.

And then it occurred to me that a frog, after all, is an "amphibian" and that of course, great blue herons eat them. What if the heron didn't eat the fish, which are so cunning and crafty, as opposed to the frogs, which are oddly guileless—or perhaps they can just distinguish a friend, as some will let me stroke their snout, eyes closed, seeming to fall asleep in the sun. It is such a conundrum, loving these creatures that kill and eat each other in that never-ending cycle of life and death in the Catskills—little brown bats, themselves now threatened, and the increasingly rare, astonishingly beautiful Luna moth, which flies at night, lured by the security light that also attracts the bats; dragonflies, those glittering flying jewels, rapacious predators of other insects, and butterflies, their wings pulsing with color as they alight on wildflowers, so easily snatched; red fox and the hapless Eastern cottontail rabbit; bobcat and white-tail deer, especially in winter, when prey is scarce, and fawn are always vulnerable, with coyotes on the mountain; hawks and songbirds, and on and on.

I was still delighted to have the heron in residence, but when it finally departed, autumn's last leave-taking—its great wings, spanning about six feet, beating slowly, deeply, as it started south—I was somewhat relieved. Almost reluctantly I made my way up to the pond. I had once found the skeletal remains of a frog there—a pair of back legs, to be exact—which I had blamed on the fish. I was dreading finding another such souvenir, dreading having to blame the wondrous great blue heron. But as I walked around the pond, the day warm for late October, I was heartened by a series of plopping sounds, as several green frogs jumped into the water as I passed. No largemouth bass were visible but that meant nothing—those devils could be anywhere in the pond's still-warm waters, their muddy green coloration providing a perfect camouflage. But when they didn't appear to mock me, peering up at me from just below the surface with their greedy eyes, I imagined that the heron's very presence had chastised them, driving the fish into the pond's shadowy depths, repentant, rueful and full of regret, sorry for their sins against frogs.

But it is only winter that intimidates the largemouth bass, the cold compelling them to seek the pond's deepest water in search of warmer temperatures, where they enter a sort of torpor, which stills even their voracious appetite (though they will still hunt, if given the chance). Green frogs huddle there too, partially dug into mud or hibernating on the bottom, seeking shelter from their predators preferably under a log but also under decomposing aquatic plants, sticks or fallen leaves. Through their skin they take in oxygen trapped below the ice, which also sustains their sullen nemesis, the largemouth bass.

Standing at the pond in winter, its surface frozen and covered in snow, as smooth as its waters on a mild day, the once luxurious cattails, which had grown green and lithe and tall, gracious sentinels, now broken brown husks edging the ice, their soft summer whisper, a rough rasp in the bitter wind, I wonder at this—that even in the Catskills, which can be so harsh, so inimical to survival, life may be slowed but is never truly extinguished.

CHAPTER SEVEN

Kaatskill Kaats

(For John James Audubon)

O LD-TIMERS IN THE CATSKILLS STILL REFER TO THEIR ROCKY
high hills as the "Blue Mountains." It is readily apparent why:
The mountains, which have reforested and are as green as
they were when the first Europeans glimpsed them in the early seven-
teenth century, seen through the haze of a sunny summer day, appear
an ardent blue. The British, who followed the Dutch into what would
become New York State early in the eighteenth century, seem to have
favored this name, as much as because it obviated the overtly Dutch influence as for its descriptive element.

In 1655, in the Visscher map of New Netherland, a creek called
"Catskill" appeared, which flows for some thirty-seven miles through
the region before it enters the Hudson River at what is now called the
village of Catskill, which also draws its name from the creek. In 1659, a
phrase in a letter written by the directors of the Dutch West India Company to Governor Peter Stuyvesant was the first to codify the range as
the "Catskil Mountains."

Still, it wasn't until the mid-nineteenth century that the name
"Catskills"—as designating both the mountains and this region—stuck.
Despite the best intentions of British officials and their colonists seeking
to Anglicize the place names of their new home, the Dutch-derived
"Catskill" emerged the winner. In large part, this was due to Washington

Irving, that singer of Catskill tales, who helped enshrine the name "Catskills" with his 1819 short story, "Rip Van Winkle: A Legend of the Kaatskill Mountains." It told the tale of a Dutch-American villager living at the time of the American Revolution (still near and dear in the memory of readers) who, beguiled by the ghosts of Henry Hudson's crew, naps for twenty years in the wilderness. Irving (who later admitted "when I wrote this story, I had never been on the Catskills") imparted mystery, even a sense of magic, to his setting, the Catskill Mountains, where convivial ghosts wandered, offering passersby moonshine that lulls men to sleep for a generation.

"Catskills" certainly seems a more fitting moniker than the smooth, rather prosaic "Blue Mountains" (even if that name was meant to resonate with the neighboring "Green Mountains" of Vermont and the "White Mountains" of New Hampshire), its rough-hewn consonantal sound evoking the craggy peaks and plunging cloves celebrated by Asher B. Durand, Sanford Robinson Gifford, and Thomas Cole in their Hudson River School paintings of the mid-nineteenth century. Their oils on canvas depicted not only America's first great wilderness—the Catskill Mountains, still pristine and as yet unexploited—but helped form the young country's national identity, as a place of wild, rugged nature, so different from the manicured, tamed European landscapes of their forebears. Today, all that remains of the "Blue Mountains" as a place in the Catskills is the hamlet of Blue Mountain and some namesake roads and businesses catering to outdoor Catskill activities.

But what is the origin of "Catskill"—or "Kaatskill"—the term that shares immortality with perhaps literature's most famous and endearing malingerer (after all, he slept through a revolution and woke to a world without his nagging wife), the notorious Rip Van Winkle? "Kaat" is a variant of "kat," Old Dutch for cat, and "kill" was the Dutch of the day for a body of water, hence creek. The most popular and enduring explanation for the derivation of Catskills, is that a wild cat— "wilde cat" in Dutch—was seen frequenting the banks of Catskill Creek.

The "wild cat" in question was most probably what we call today the bobcat, a medium-sized feline with a signature bobbed tail, spotted brownish coat, and ears with black tufts. But bobcats are elusive, rarely seen by humans, even today, and there is no record of bobcats patrolling the Hudson River or its tributaries. Mountain lions, often called in the Catskills "catamounts" (from Old English,

meaning "cat of the mountain"), which were also known to be in the area when the Dutch arrived, is another plausible feline explanation (though chances are they were less common). Of several other theories, perhaps the most appealing from a literary perspective, is that the Catskills were named for the Dutch poet Jacob Cats, a not uncommon naming practice of the time. According to historian T. Morris Longstreth, "At the very time that Hendrik Hudson was eating roast dog with his red-faced hosts near the outlet of the brook that was to be Cats'kill, Mr. Cats was penning amatory emblems behind his native dikes."

Of the numerous explanations, a few of which are mentioned here, the most popular, which has insinuated itself into most sources, is that the name "Catskills" derives from that "American wildcat," as the renowned naturalist and painter John James Audubon called it, known today as the bobcat. It is the only wild cat certain to endure in breeding populations in the Catskills, even if sightings are so rare that many Catskillers have never seen one, even after living a lifetime here. To wit, when my neighbor on the mountain, who keeps honeybees and is thoroughly conversant with local fauna, reported he had spied a bobcat picking its way through the snow on a winter day, I was skeptical, thinking he had probably seen a barn cat, its coat thick and furry for winter. But when I saw what I assumed to be the same animal the next afternoon in the upper meadow pouncing on some small prey, a bobcat for sure, I realized something about this wild cat as well as about observing wildlife—that critters will always surprise you, and that however farfetched people's claims may seem, especially those of the residents of this rocky terrain, who, like the critters themselves, are always watching, they cannot be dismissed out of hand.

EVEN IF DUTCH SETTLERS SAW THE BOBCAT INFREQUENTLY AT WHAT would become known as "Catskill Creek," is it so farfetched to think they would name it for this "wilde cat"? The bobcat is native to North America and would have been, for the Dutch, a new creature. The Eurasian lynx—a larger cat, also with tufted ears and a stubby tail—would have already been extinct in the Netherlands for centuries, loving, as it does, the deep snows of the boreal forest. Sometimes, I think, we forget the persuasiveness of wonder, the wonder Europeans must have felt at seeing these critters of the New World, just as those left behind must have wondered at the monumental landscapes expressed in the paintings of the Hudson River School. What perhaps surprises me is that chances are, settlers would have seen this wild cat at the creek during the day, as bobcats are primarily crepuscular, favoring the hours of dawn and dusk when their prey is most active and when the shadows allow them, wearing their spotted coats, to become shadows themselves.

But again, perhaps we also forget the impact we have had on wild critters, including their behavior. Black bears, both here in the Catskills and in other places where they have learned to abide people, have become more active at night than in the daytime, primarily to avoid us. Wildlife biologist J. David Henry reported that in Saskatchewan's Prince Albert National Park, a 1,500 square mile wildlife sanctuary where he conducted extensive fieldwork on the red fox, usually as elusive as flickering light, fox appeared quite openly in his presence, having no fear of—and perhaps having no reason to fear—humans. Bear, fox and bobcats too certainly would have been aware of local Native Americans who hunted them for their fur, and, in the case of the bear, for meat. But it was Europeans who brought on the wholesale slaughter of wild animals in the Catskills to capitalize on the fur trade, as well as the technology to do so.

I would learn that my neighbor's sighting —and then my own— of the bobcat on a winter afternoon was not really that exceptional. The bobcat adjusts its behavior, becoming more diurnal in winter because the small mammals it primarily hunts—rabbits, mice, and voles—are more active during the warmer daytime hours. But there was no way that I could account for my first close look at a bobcat, or explain away its behavior, when it sauntered by me that sunny day in June . . .

It just appeared, this most cryptic of critters, notorious for its secrecy, the one animal I was told when I moved to the mountain "that sees you, but that you will never see." But there it was, walking idly by the back deck, completely unhurried, mere steps away from where I was standing at the screen door, seemingly invisible to this critter that sees everything. I was certain it was a bobcat, not some raggedy tiger tom with torn ears and matted fur that visited occasionally, eager to ogle the thistle seed feeders I hung in summer for the goldfinches, an interloper I would chase away. What identified the bobcat and made it instantly recognizable was the black tufts of hair at the tip of each pointed ear and its short, bobbed tail—this cat's "bob" I estimated at four to five inches— also tipped in black, which gives this species of wild felid its name. Or was it the way the bobcat walked, I wondered, that had earned the cat its name, noticing its muscular hind legs that gave the bobcat its "bobbing" gait.

But there was nothing awkward in the way this animal moved. On the contrary, it appeared jaunty, insouciant even, striding easily, clearly comfortable in its surroundings which, I now realized, was not just my territory but part of the bobcat's as well. I had seen it emerge at the edge of the woods to cut across my property, clearly a path familiar to the cat, and I marveled at how it had evaded me, who is always watching for the critters that watch for me.

But cats, as anyone who keeps them ("own" is not an operative word when it comes to felines) knows—those "domestic shorthairs" that are at heart always feral, a contrary species that does what it wants regardless of our pleas, and still manages to elicit our affection—always surprise. This bobcat had stunned me by appearing so close to the house, shortly after noon. Now it startled me again by sitting down at the foot of the back deck, just below the bottom step, pausing, incongruously it seemed, between those human contrivances, the deck and the carport.

The cat was sitting on its haunches in profile, facing the western woods, where I assumed it was heading, picking this path as a shortcut instead of skirting the edge of the woods where it meets the meadow, staying under cover. What could this bobcat be thinking, to be so bold in broad daylight? As if to answer my question, or perhaps dismiss it, the bobcat began to lick its paw then swiped at its ear, a half-hearted attempt at grooming. Then it simply sat there, staring into the distance. But its body wasn't taut, as if sighting prey—the bobcat just seemed to be taking in the view, enjoying the day.

The bobcat depends primarily on its sight and hearing to hunt, its ear tufts helping to corral sounds, especially of scurrying critters. Its sense of smell, though good, cannot compare to the dog's and is mostly used in "social behavior" to communicate with and identify other bobcats. If I stayed silent and didn't move, as still as a bobcat waiting for prey, it just might not notice me, I reasoned. What struck me from the vantage of my blind was the litheness of the bobcat. It had a kind of coiled grace, as if it was ready to spring, even in this brief moment of rest. I watched the critter, not moving, refusing to breathe, not wanting to alert it that a predator was watching it.

The bobcat is typically described as being twice the size of a domestic cat. If one considers the "stats," as posited by the New York State

Department of Environmental Conservation for the bobcat average—weighing about twenty pounds, standing at least sixteen inches at the shoulder, body length some thirty inches or so, standards this cat seemed to meet—that description fits. But statistics cannot measure the strength, the muscularity of this creature, its very solidity, a pit-bull among felines. And indeed, the bobcat is known for its ferocity, the culprit of that frontier tribute to a fighting man's courage and strength: "He can lick his weight in wildcats." The bobcat looks like an athlete, a halfback that can sprint, leap and tackle.

When I spotted the bobcat last winter (could this be the same one?), pouncing on what I suspected was a field mouse or some other hapless critter, I noticed that its coat appeared gray. My research revealed that, in winter, the bobcat's fur is longer for warmth and also turns a tawny gray. This helps the bobcat blend into the landscape as it hunts, and the white-tailed deer (which it will take, especially in winter, a meal that will last for days) counters with its own grayish camouflage. But today the bobcat was in its summer finery. Its coat was cinnamon-brown with a russet cast. I could clearly see the fur's spotted patterning, dark streaks that, in the fading light of the twilight forest, would render the bobcat a shifting shade. The bobcat's short, thick coat, rich in color, immaculately kept, seemed to gleam in the bright afternoon light. It was a beautiful creature, this bobcat. But there is a price for that beauty: The bobcat continues to be hunted—and trapped—for its pelt, which makes humans the bobcat's enemy, and in the Catskills, its one true predator.

Abruptly, the bobcat stood up and resumed its leisurely but purposeful pace. I slipped out onto the deck as quietly as I could, not wanting to lose sight of the cat. I was close enough to see the white patches on the back of its ears; another bobcat diagnostic, which some observers believe enables kittens to keep track of their mother traveling through high grass. Healthy bobcats, though fierce—not even bobcat kittens should be handled—do not attack humans. I knew I had no reason to fear this cat. Indeed, I was curious to see how it would react to being caught out in the open by its primary predator. I could not resist calling to it, and even relished the silliness of addressing the Catskill wildcat as if it were a docile domestic feline:

"Here, kitty, kitty, kitty. Here, kitty, kitty, kitty . . ."

The bobcat whipped around, stared at me with what I imagined was indignation. Its ruff—long, luxurious hairs that adorn both cheeks—and its brush of copious whiskers made the bobcat look utterly alluring, plush and savage at the same time. I was smitten. Then the bobcat bounded away, fading into the tall grass where the meadow meets the woods. I waited for some time, hoping to catch another glimpse, a ripple of movement. But the bobcat did not give itself away and, I suspected, it would never give itself away again.

I WAS RELUCTANT TO TELL ANYONE ABOUT MY ENCOUNTER WITH THE bobcat. Frankly, I was afraid people would think I was making it up, my own Catskill tall tale. Most of the folks I knew had never seen a bobcat in the wild. I had stopped trying to fathom the bobcat's behavior, why it had appeared out in the open, during the day and so near the house. I did know that the bobcat must have known I was around. Wild animals always know when people are present, and just an hour before it appeared I was sitting with friends on the back deck, talking and laughing. But it is impossible to know what—and how—a wild animal thinks, why it does what it does, especially if that behavior is not keyed to its survival. Perhaps it was as simple as the bobcat taking a stroll, savoring a pretty June day, surveying its environs.

I did tell Marvin, an accomplished bow hunter, who now prefers "just to watch wild animals." He still tracks critters but "for the fun of it," and has told me that the bobcat is the most challenging to follow.

I told him my tale over a cup of coffee, sitting at a local diner that has faded tan wallpaper adorned with black bears. It sounded, I thought, improbable, even to my ears. When I confessed that I had called to the bobcat "Here, kitty, kitty, kitty," Marvin frowned. Had I offended him, the hunter and tracker, who so respected wildlife?

Finally, when he didn't speak, I asked him, "Well, what do you think?"

Marvin took a swig of his cold black coffee, his ever-present tooth-pick still in place (I was always afraid he would swallow it). "Just maybe," he began, then paused, as if considering, "I think just maybe you saw yourself a ghost."

"GHOST CAT" IS A TERM THAT APTLY DESCRIBES THE BOBCAT, AND IS also a nickname for those other wildcats that once roamed the High Peaks, the Canada lynx and the mountain lion. It was the cats' solitary nature, as well as their ability to blend into their surroundings, silent wraiths that could move through the forest without snapping a twig, that earned them this sobriquet. Native Americans, who hunted the cats for food and fur, and in some tribes for the ornamental value of their teeth and claws, no doubt revered them as predators. But the ghost cats remained largely elusive, keeping their secrets even from these skillful hunters. To wit, wolves are featured twenty times more in tribal legends than mountain lions, and bobcats and lynx are even less in evidence.

Of the three species of wild cats, only the bobcat endures in the Catskills today. Both the lynx and mountain lion are considered "extirpated"—an ugly-sounding word that means extinct locally, though the species may exist in other environs—in New York State. The bobcat has survived because it is adaptable. It is small enough to secret itself in a backyard woods, its spotted coat camouflaging it from human view. The bobcat, as opposed to the fussy felines that deign to live with us, is *not* a picky eater. Its wide-ranging diet has been known to include forty species of small mammals, birds, and deer. This undiscerning palate has enabled the bobcat to thrive in habitats as diverse as woods and wetlands, desert and forest edge, as well as the rocky hillsides of the Catskills, where a cat can readily find cover in the thick, brushy understory. Perhaps the key to the bobcat's success is that it can tolerate the presence of people, at least on its own terms, as it navigates around farmhouses and even suburban developments, usually at dawn or dusk—a phantom, if seen at all, caught out of the corner of an eye.

As the bobcat has survived due to its adaptability, the lynx is a success of specialization. The lynx, of the same genus as the bobcat but a different species, looks like its big brother, sharing a similar appearance. But their differences are crucial to the lynx's ability to thrive in its preferred habitat, the spruce forests of Canada, Alaska, and several northern U.S. border states. The lynx is grayer than the bobcat (*Lynx rufus*, its scientific name, indicating the bobcat's darker coloration), consistent with the terrain it travels, the deep snows of the boreal forest, and its fur, tinged in russet, is thicker and longer, to protect the lynx in bitter cold. Its cheek ruffs are shaggier than the bobcat's, giving them the appearance of mutton chops, which warm the lynx's face like a mountain man's beard, and its black ear tufts are longer too, to capture the soft tread of prey in the snow. The lynx also sports the iconic bobbed tail, though it is shorter, perhaps making it less vulnerable to frostbite.

The lynx, a lanky cat, is only slightly bigger than the bobcat but it appears taller. Its legs are long and the back are longer than the front— "big jackrabbit hind legs," an example of the predator imitating its prey, the snowshoe hare, whose population fluctuations affect the lynx's own, so tied together are these two species. But what really separates the lynx from the bobcat and marks it as a creature of its range, is its feet, which can best be described as furry snowshoes (greater than three-and-a-half inches long and wide, about twice the size of the bobcat's), allowing the lynx to glide swiftly over snow.

The snowy slopes the lynx is so well suited to, in a sense, insulated it from competition. Those other wildcats, the bobcat and cougar, don't fare well in deep snow; the bobcat is ill-suited for snow because of its shorter legs, and both cats lack the lynx's formidable feet. It has also been suggested that this isolation may explain the lynx's less aggressive tem-

perament. It has not had to battle, as the bobcat has, mountain
lions, wolves, and even coyotes,
where they occur together in the
same range. And where the bobcat and lynx do coincide, mostly
along the Canadian border, the
smaller bobcat usually gets the
best of the scrap. But human intervention has broken the lynx's
isolation, certainly in the Lower
48, with habitat loss due to forest
fragmentation, the result of log-

ging, and everywhere a warming climate is melting deep snow, inviting
cougars and bobcats into its territory, driving the lynx to higher terrain.
The generalist, more versatile bobcat (which can live almost anywhere
and eats almost anything, a tough customer that can pick its way around
people) seems best equipped to survive in our rapidly changing world.

By the late-nineteenth century, the lynx and the mountain lion had
disappeared from New York State due to hunting, trapping and, in the
case of the cougar, a bounty placed on the head of what was perceived as
a dangerous predator and threat to livestock. It is generally accepted that
the lynx included the Adirondacks and perhaps even the High Peaks
Catskills in its range. The bobcat is even blamed, at least in part, for the
lynx's disappearance. When the Adirondacks, favored by lynx for their
spruce tree forests and plentiful supply of snowshoe hares, were exten-
sively logged in the 1800s, so the theory goes, this opened up land for
white-tailed deer and then the bobcat, which fed on the deer and routed
the milder lynx. Since these forests have regenerated, thousands of acres
of which are protected in the Adirondack Park, the case was made for
the lynx to be returned to their former home. Over eighty lynxes, some
of them pregnant females, were released in the Adirondacks between

1989-1992 by the SUNY Department of Environmental and Forest Biology for a study funded by New York State. All of the lynx were radio-collared, allowing them to be tracked as far away as Canada, Pennsylvania and New Hampshire (unfortunately, a substantial number were killed in traffic). The lynx dispersed far and wide but they did not colonize the Adirondacks, giving credence to the NYS DEC's assertion that we don't really know if there ever was a viable breeding population of lynx in New York. Perhaps their numbers were simply augmented by lynx passing through.

Dr. Rainer H. Brock, who headed the lynx restoration effort, offered this rationale for the project: "In a society and world that is very complex, there is a need more than ever to have a natural world that inspires." The lynx, for Brock a symbol of wilderness, "is an unusual animal that can inspire us." Lynx are spotted fairly regularly in the Catskills; several friends of mine swear they have seen them. Most of these are probably bobcats, though certainly some could be a traveling lynx. What fascinates me is the need, even the will, to see the lynx. The sight of it is like peering into the past, an untamed and untrammeled world, a world in perfect balance, which was already beginning to be lost, transformed into myth, when the painters of the Hudson River School were celebrating the Catskills.

SOON AFTER I MOVED TO THE CATSKILLS, I STARTED HEARING ABOUT mountain lions. These stories were told matter-of-factly, even casually, often with a shrug. Mountain lions, it seemed, were taken for granted—they were *here*. "Here" wasn't just the Eastern Catskills, which feature thirty-five peaks of 3,500 feet or more, their flanks forested with stands of balsam firs, a resilient wilderness, where rugged cliffs and steep cloves would offer a formidable predator safe haven; "here" was the Western Catskills, specifically, the rolling foothills of Delaware County, which nature has overtaken since the cessation of industry, returning white-tailed deer and black bear to its once-again verdant slopes.

The majority of these sightings seem to take place in the vicinity of the Cannonsville and Pepacton Reservoirs, whose banks, along with significant buffer lands, are unbroken by roads or any development that could trammel the quality of New York City's drinking water. Some recreational use—carefully regulated boating and hunting, a sop to locals who still harbor resentment over the villages that were razed to create these inland seas in the '50s and '60s—is allowed on reservoir lands. But they are, in essence, wildlife preserves, especially in the upper reaches, with little, if any, human encroachment. It didn't seem so far-fetched to me that the "Catskill cougar," as it is locally known, could be in residence.

Indeed, the first report of a mountain lion I received was from a woman who ran an antique shop in the village of Andes. I was admiring a small bronze cast of a cougar when she volunteered, "I have one of those, a mountain lion. I see it every morning, passing through my property, on its way to hunt, I figure." She lived on the Tremperskill Stream, which meanders its way into the Pepacton, in the DEP (NYC Department of Environmental Protection) watershed preserve. On another occasion, Clem, who had come to fix my propane stove, told me about the "little black panthers" that arrived every day at his creek, which he claimed to hand-feed.

Perhaps more persuasive, to a skeptic, are the accounts of seasoned observers, officers of the DEP police, who are charged with patrolling the reservoirs day and night, several of whom have given detailed descriptions of mountain lions in this area since 2000. One such sighting occurred outside nearby Walton, a village on the West Branch of the Delaware River, in 2004. What convinced Marshall Vandermark, also a DEP officer, was the critter's long tail and "bland" coloration. Like most who have glimpsed the "Catskill cougar," Vandermark, an experienced hunter, was adamant: "If I'm positive it's a mountain lion, then it's a mountain lion." Do these wild cats really exist here, or is the mountain lion a rural legend, the Catskill Bigfoot?

What makes people "positive" that they have spotted the mountain lion is its long, heavy tail, which measures twenty-six to thirty-two inches long. Its consistent tawny color is also determinative, true to its scientific name, *Puma concolor,* which means "cat of one color." Then there is the mountain lion's size—it weighs, on average, 180 pounds (with a range of eighty to 225 pounds), and its body length, including its trademark tail, is five to nine feet. The mountain lion is five times the size of a bobcat, the animal that "experts" most often try to convince true believers they have seen. In 2011, the U.S. Fish and Wildlife Service finally declared the cougar extinct in the Northeast, and the big cat, as extirpated, was removed from the endangered species list in 2015. Ironically, USFWS's listing of the Eastern cougar as "endangered" in 1973, even though a remnant population was not known, perhaps has led to the persistent sightings of mountain lions in the Catskills, Adirondacks, and many other locations in upstate New York, as well as the conviction that the existence of these wild cats is being "covered up" by the powers-that-be.

It is perhaps understandable, the reluctance of those who insist they have seen this critter, to accept bobcats and house cats, even dogs, as the explanation. The mountain lion is mythical, revered for its stealth, the liquid grace with which it moves, pouring itself over rocks, insinuating itself into the landscape, its ability to kill, quickly and efficiently, with one brutal bite, an executioner among predators. It stalks its prey, a deer, preferably, and then ambushes it, leaping onto its back, its sharp, saber-like teeth severing the spine at the neck. "Silent death" is one of the mountain lion's well-earned nicknames.

In the Catskills, "catamount" was one of the most popular names for the mountain lion, immortalized today in the names of resorts, restaurants, and even a museum, though the most fitting tribute is in the Adirondacks: Mt. Catamount. Of all the mountain lion's forty-two names in English, the most familiar is "cougar," which derives from the Tupi word for "false deer." The Tupis, an indigenous people of the Amazonian rainforest, seemed to grasp a very modern scientific concept, that a "predator evolves to blend into the same habitat as its chief prey," a reference, in this case, to both these critters' tawny coats. The balance between predator and prey is key to a healthy ecosystem. Left to forage unchecked, deer can decimate local flora. Bobcat prey on deer, as can coyotes, but in the Catskills, the primary role of top predator is played by human hunters. The white-tailed deer is also a principal in the ongoing drama of cougar sightings.

The presence of mountain lions had been reported in upstate New York for decades. But in 2001, a rumor began to circulate that the DEC (New York State Department of Environmental Conservation) had released eleven cougars in Southwest Albany County near the banks of the Hudson River, where Europeans probably first saw the New World lion, and then in 2007-2008, released more cougars (and supposedly gray wolves!) in Columbia County. The DEC denied this, but their dismissiveness of the claims of "credible witnesses" did little to staunch these stories. In the Catskills, suspicion quickly turned to distrust of the DEC. Its sister agency, the DEP, is charged with overseeing the reservoirs that flooded nine hamlets, displacing the living and the dead, which were dug up and moved. It is fair to say that governmental agencies, and their assurances, are looked at askance in the Catskills. A conspiracy, man-eating wildcats, and of course, a cover-up—what could be more alluring?

The DEC strenuously objected, issuing this disclaimer: "The DEC has never released cougars, despite what you may hear to the contrary." On the same website, "Eastern Cougar Sightings," it has published "false images" of mountain lions claimed to be local cats, which are, in fact, the Catskill cougar's western cousin. The DEC does concede there have been "a few isolated sightings," probably of cougars that escaped from licensed (or even unlicensed) facilities, or of an occasional wandering wild cat.

The rationale for these supposed releases is that the DEC wanted to reduce the overpopulation of deer, which damages agriculture and contributes to the decline of biodiversity. A less elegant theory is that the DEC is in cahoots with the insurance industry, which every year takes a substantial financial hit for vehicular collisions with deer. DEC wildlife biologist Nancy Heaslip, who finds such speculations daunting—"I don't know how these rumors start"—points out that the "DEC would never release a species without public approval," a "rigorous and heavily regulated process of open debate, review and public commentary." What we do know is that almost certainly, no self-sustaining, native population of mountain lions exists in New York State. There just isn't a body of evidence—tracks, scat, hair, even the carcasses of deer, whose wounds would reveal cougar bite marks, let alone definitive photographs or videos—that offers conclusive proof.

But as rumors persist, there is a growing sense that the mountain lion could—and should—be restored, certainly to the Adirondacks, which would give it room to roam farther away from people (though the Catskills would proffer them an endless supply of deer), but the feasibility of this is much debated. There is even some indication that in time the cougars' return might happen naturally, without human intervention. In 2010, a mountain lion was spotted in Lake George, a village in the Adirondacks, and subsequently huge tracks were found in the snow, definitively a cougar's, along with hair whose DNA linked the cat to mountain lions living in the Black Hills of South Dakota. The cougar, a three-year-old male, which had trekked over a thousand miles in its journey to find a mate, sadly, was killed by a car near Milford, Connecticut in June 2011. Perhaps, in the coming years, other mountain lions looking for love or adventure on their own amazing journey might settle in the Catskills or Adirondacks and start their own families. Certainly, in some quarters, these big cats would be most welcome. Top predators, such as the mountain lion, are needed to ensure the health of the environment.

"We have forests that have lost the top and bottom of the food chain," says Michael Robinson of the Center for Biological Diversity. "It should be a clarion call to recover pumas and all our apex predators to sustainable levels to help rebalance a world that is out of kilter."

My road down the mountain has always been a place where critters congregate. There is very little activity on it. Most of the time I am the only one here, and the *chug-chug* of the mail truck is the sole sound that breaks the silence. I really shouldn't be surprised. Critters take advantage of any trail I cut through the woods, and in winter when I shovel, the cleared path is soon stamped with turkey tracks and the hoof prints of white-tailed deer.

Woodchucks live at the top of the road; I often see them scampering across it. I am always watching for deer. They pop out of the trees and, panicked by the car, charge in front of it. I have stopped to let a porcupine pass, and to shoo away a sluggish snapping turtle, thankfully small, sunning itself on a crisp autumn morning. In the high grass edging the road, a red fox may appear—a twilight apparition—hunting rabbits and voles. Chipmunks are everywhere. They seem to delight in racing back and forth, a dangerous entertainment, just as I approach.

I head down the road slowly, and at night, creep up it, high beams on, which sometimes catch the eerie glow of light reflected from a critter's eyes. Afternoon is the time I am least apt to see animals but still, I am cautious. So I wasn't surprised to see two cats playing in the road that August day. I assumed they were siblings because they looked identical, tigers with warm brownish coats and striking black markings. They were young, I guessed, from how they cavorted, running, bumping, jumping, then wrestling each other to the ground. One held the other down with a foreleg, paw firmly planted on its chest, just as Hannah Bean, my pearly gray, does to her calico "sister," Ellie Bellie. I can see the wildness, even in my house cats, when they play, their seemingly good-natured roughhousing honing the predator's edge.

But these two cats were larger than house cats, sturdier; these two cats, I realized, had bobbed tails—they were bobcat twins! Abruptly the twins disentangled, scrambled to their feet, and looked straight at me. I had stopped the car to watch them tussle and now sat there, thrilled at the sight of the young wildcats. Then they bolted from the road, in perfect unison, as if they had been summoned. I watched them scramble up the embankment to a low hillock where another bobcat, bigger than the twins, which I presumed was their mother, was waiting. She looked at me intently, without deference, appraising me. Her eyes were yellow and, I noticed, round, their shape a bobcat signature. They were standing close together, as if posing for a family portrait. I blinked, and the bobcats were gone, vanished into the thicket of wild rose.

I ESTIMATED THE TWINS TO BE FIVE MONTHS OLD OR SO—THAT'S about the age bobcat kittens start hunting with their mother. I had seen them in August, and as bobcats start being born in the Catskills in early spring, mid-March and April, this seemed a reasonable estimate. I knew they would separate from her in a few months, before winter set in, when their mother would breed again. I was—and still am—amazed at my adventures with bobcats. I had seen a bobcat hunt, which I was told was exceedingly rare to see any wild cat do, and I'd witnessed, in the wild, bobcats at play. These were incalculable gifts that, truly, made me want very little else.

Bobcats, with their shorter legs, do not favor deep snow—even a foot of the white stuff is difficult for them to navigate. They will often wait out a winter storm under a rocky ledge or seek shelter under a stand of evergreens, whose thick, burly boughs shield them from heavy accumulations. Though bobcats can go for days without eating, if they do have to travel, they may use a makeshift trail of logs and rocks, jumping from one to the other. They are adept athletes and very canny, overcoming many obstacles to their survival.

That winter the snow started to fly at the end of December. It piled up gradually, over a foot in a few days. I shoveled past the driveway, which had already been plowed, past Woodchuck Rock into the western woods, so that I could access the bird feeders I hung there. Then I dug an even longer path in the opposite direction, all the way to the Cut, where I scattered apples and cracked corn for any enterprising critters. In the evening, the snow started up again. Around midnight the motion detector light on the barn triggered, illuminating the driveway. Usually that means a deer is passing by. But when I looked out, I saw first one bobcat heading for the house, then the other. They sat down beside each other directly below the window where I was watching, greeting each other with head butts. Then they got up and moved off together in the direction of the back deck. I couldn't be sure, but I suspected that the twins, now fully grown, these ever-adaptable bobcats, were taking shelter from the storm under the "ledge" provided by my deck.

The next morning they were gone but in the fresh snow I found bobcat tracks, easy to identify, featuring four toes and a heel pad, about two inches in diameter. I followed the tracks into the Cut as far as I could but they disappeared abruptly. It was a sunny day, mild for late December, the temperature warming into the forties. By afternoon the tracks had melted away, just as the bobcats, ghost cats of the Catskills, had dissolved into the forest.

CHAPTER EIGHT
Tracking Winter

(For Annie Dillard)

NOVEMBER'S RELENTLESS MONOCHROME—ITS HUES OF DARKening brown—has settled over the mountain. Only the stubborn oaks still hold onto their leaves, reluctant to cede to winter. The oaks are the last to change color in October, as they are the last to leaf in May—so perhaps it is in their character to be recalcitrant, feeling no shame that their long, lobed leaves, once supple, a rich deep autumn bronze, are now brittle and curled. The green that is summer here, that infuses the short, glorious growing season of the Catskills with life, has been distilled into mere spikes of color, stands of evergreens, pines and spruces especially, scattered among the maples, beeches, and aspens, sleeping now. The high grasses of the meadows, pale in spring, verdant in summer, ripening to amber in early autumn, seem almost transparent, ghosts of their former self, lifeless save for the sway of the wind.

When the first snow comes, which can be as early as October, it awakens this landscape, so bereft, edging it with a new beauty. Snow is a sculptural medium, shaping in white as pure as Parian marble everything it alights on—a half-buried boulder resembles, at twilight, a slumbering beast; the rugged bark of a mighty oak, the "old man of the meadow," is transformed into a delicate frosty latticework; and in the Catskills, the scars left by the bluestone quarries, made visible on moun-

tainsides when the leaves fall, are once again healed. Even sound is assuaged by snow, becomes less sharp, gentler to the ear, as light is made
more intense, expanding the eye, the low winter sun suddenly brighter.
The genius of snow, why it pierces us, is that it changes our perceptions
as well as the world. Nothing else can do that. When I wake to new-
fallen snow, especially the first of the season, when yesterday the ground
was bare, the hillsides barren, all now covered in glistening white,
however fleeting the feeling, I am made new.

WINTER, WHICH CONCEALS SO MUCH OF LIFE, ALSO REVEALS; IT OFFERS
us the tracks of critters as compensation, as a way into their now secret,
silent world. Tracking, for some, sharpens them for the hunt, gins up
the predator. But I track animals to be close to them, to see some small
sign, to catch a glimpse, now that winter has robbed me of the sights of
summer: a porcupine waddling furiously along the edge of the upper
meadow at dusk, on its way to the pond; a red fox appearing—and disappearing—in the early morning mist, ever elusive; a coyote trotting
down the mountain path, as purposefully as if it were heading home.
What I miss in winter is the animals and their stories, which is what
their tracks, made in ephemeral, reliable snow, capturing their nights
and days, tell me.

There are two critters I can track with my eyes in winter, at least
until the drifts get so deep that navigating becomes difficult even for
them, these Catskill survivors—the wild turkey and the white-tailed
deer. Their tracks are the first that anyone learns here, they are so
prevalent, and as common as both animals are now in the Catskills,
seeing them always evokes for me a sense of wildness and a kinship
with earlier settlers. They tracked turkeys, too, for food in snows like
this one, coming so late, at the end of December (though that, I knew,
presaged nothing, not a mild winter or a bitter one—weather is so mercurial here, and unpredictable). This first snow was a good tracking
snow, measuring no more than three inches, a snow that shows off

tracks as crisp line drawings, turning hooves, paws, even turkey toes, into minor works of art.

Something about turkey tracks strikes me as prehistoric, conjuring up images of charging velociraptors—perhaps it's also the way turkeys run, their neck extended, looking even longer, their narrow head, too small for their body, jutting forward, as they race each other to the feeders. Their eerily reptilian feet feature three toes (the fourth, at the back of the foot, does not track), which birds commonly have. But the turkeys' toes are longer, four inches or so, with a pebbly texture that prints in snow, along with a rakish claw at the end of each digit. This morning, the mountain's winter flock (twenty-five, at last count) must have traipsed through—the pristine snow is stamped with a frenzy of turkey feet—eager to scrounge the mixed seed I put out for the juncos and other ground-feeding birds, which are shy of the feeders. They are survivors, these dinosaurs of the woods; that is their story. The wild turkey has rebounded from being hunted to the brink of extinction in the Catskills and the rest of its North American range. And though I may scold it for its gluttony—earning in response indignant gobbles—I agree with Benjamin Franklin who, though he characterized it as "a little vain & silly," called the turkey a "bird of courage" and a "true original Native of America," and submit that for its resilience alone, how it has triumphed, the wild turkey would make a distinguished national emblem.

The birdfeeders at the edge of the western woods, after it has snowed, are an excellent place to track. I have found signs there of opossum, shy and unassuming, the drag of its long, skinny tail, its pink humanlike feet, giving it away, along with the tracks of raccoons, squirrels, rabbits, and meadow voles. Today, though, I suspect the turkeys have obliterated any tracks in their undignified dash for food, scratching like chickens for stray seeds. So instead I visit the cedar shake shed, which in summer is a resort for the mountain's smaller denizens, mice and chipmunks especially. The chipmunks, clever and quick, would be denned up by now, sleeping with their stores of nuts and sunflower seeds, which

I feed them in fall. Their cheeks stuffed, they bound over the open land, tail straight up, wary of hawks, to their rock lodges at the forest edge. But today my inspection reveals the trail of another critter, the white-footed mouse. To give a sense of this diminutive mammal, its hind foot is three-fourths of an inch to one inch in length—the tracks of its toes in the snow are indistinguishable, they are so tiny. What marks the white-footed mouse, what sets it apart in the world of rodentia, is its tail. It is long in proportion to the mouse's body (which is up to eight inches in length, about half of which is tail), leaving behind a thin, telltale line in the snow.

The white-footed mouse has winter gray fur (reddish-brown in summer), a white throat and belly and white hands and feet, daintily marked. It has large ears and eyes, which John Burroughs winsomely described as "full of a wild, harmless look." The mouse's tracks, at the side of the shed, seem to go nowhere. But I suspect the mouse's story. The shed is set on concrete bricks, unevenly spaced—between them are pathways providing ample room for the mice and in summer, the chipmunks, to travel. Chances are, the mouse had popped out from its labyrinth to survey the first snowy day. The tunnels make not only ideal hiding places for these rodents, ever imperiled by predators, but also storage caches for food—or perhaps it was the shed itself that this little white-foot coveted. Every spring, when I open it to the air, an acrid odor escaping, I find piles of hulled seed and other food scraps, souvenirs of the mouse's stay.

Mice must never be underestimated, despite their small stature. In addition to being highly adaptable and taking everything of ours as their own (mice will even eat soap!), they are an engineering marvel. Mice *do* have bones, regardless of the old wives' tale, but the mouse's skeleton is so flexible, including its ribcage, that if it can sneak its head into a hole, the rest of the body will follow. I witnessed this firsthand when a white-footed mouse breached the back door, rushing into the house, past the cats, those stalwart sentries, whose thumping feet alerted me to the chase.

The two cats finally cornered it but just as they were about to pounce, the mouse ascended the wall to a hole it had spied, no larger than a nickel. There as we watched, spellbound and amazed, it stuffed its entire body into the opening—the body seeming to collapse—resembling as it disappeared, a limp wad of tissue paper. This is the mouse's arête— that it can climb into the smallest

spaces to escape predators such as my poor, flummoxed felines, and to make a safe home, where it will breed legions of offspring.

On the other side of the shed, I pick up the trail of a white-tailed deer family, where they have emerged out of the woods. I have followed them since both fawns were born in June and I can see now the tracks of three different deer. The hoof is cleft, heart-shaped, narrowing to a point in front, which indicates the way the deer is traveling. The larger doe was in the middle. Her tracks, at about two and a half inches long, are slightly but identifiably larger than those of her twins. Deer are also "diagonal walkers"—they lift the front and hind leg on opposite sides at the same time, leaving behind staggered tracks, which is in evidence here. The trail leads to the lower meadow, where I can tell from the patches of kicked-up snow the deer have used their hooves—and probably their muzzle—to clear the dry grasses which have so far fed them.

Through binoculars last summer, I saw the doe give birth in the lower meadow. She was standing in the high grass, among the milkweed, the "butterfly flower" that nurtures monarchs and sustains many other species, whose fragrant pink blossoms were just beginning to bloom. When she lay down, I wasn't surprised. I have found many large whorls in the meadow grass, where the deer like to bed down for the night or rest in the afternoon. When I returned a half hour later, I was startled to see another head had appeared, smaller than the doe's, low in the grass. This time I waited, peering through the binoculars, but my vigil was only fifteen minutes. Still another deer head emerged, joining its twin in the meadow. Life is not easy here in the Catskills, especially for fawns, prey for coyotes. But on this June morning, sunny and warm, the sky blue, cloudless, with a bare ribbon of white, the day promised much.

I watched as the newborns struggled to their feet and stood, wobbly-legged, nursing as their mother cleaned them of afterbirth, licking them harshly with her tongue, imprinting her own scent on them, and catching theirs. Later, I would see her repeatedly lick their bottom to stimulate urination and defecation, crucial for their survival. I monitored the family through summer as the fawns started to lose their "camouflage," their protective white spots, shimmering like specks of sunlight in the tall grass, and grew stronger, worrying over their vulnerability to the coyote pack on the mountain. Then, in November, when hunting season began, I silently implored them to stay close, on posted land, though I could hear the ricochet of rifle fire all around.

I wasn't worried that the fawns would be shot. But hunters do shoot does, sometimes even with fawns. Only last winter, on Christmas Eve, I spied out a front window, the moonlight reflecting off the snow, another doe with her two fawns—only this deer had a shattered foreleg. I had no idea how she was standing, the bone was splintered, showing through, clearly the result of rifle fire, the recent hunting season. Seeing her with her twins, her leg dangling, I have never felt so helpless. I could only hope that she would be taken down by coyotes—a gentler, fairer, quicker death than this. I never saw the doe again, but I did see her twins who were on their own only briefly, as they were "adopted" by another deer family with two does—perhaps one an older sibling—and their own winter fawns.

It was a happy ending to the sad story of this deer family. But the image of that doe standing in the snow, illumined too sharply by the moonlight, her leg riven, I will never forget.

TODAY IS FEBRUARY 2ND, GROUNDHOG DAY—THE PRECISE MIDPOINT between the winter solstice and the vernal equinox, which many traditions considered the start of spring. The sun has not appeared in the Great Western Catskills for nearly two weeks. If Jack Kerouac, resident woodchuck, hibernating beneath his bluestone boulder, had wakened to search the stark landscape for his shadow, he would have found none. Still, I am not assuaged. Winter will not be over soon. In January, a stubborn storm that frittered for days, snowing lightly but insistently, left over a foot of snow on the mountain. Since then, despite a thaw—winter's false offering, melting snow only to refreeze it, capping it with an icy glaze—the snow has accumulated to three feet, higher in the meadows where the wind has driven it.

Looking out, I can see the deer family—the doe and her twins—in the woods at the edge of the upper meadow, which, no doubt, they have found impassable. They are huddled together in a stand of white pines, by design, I assume, as evergreens, with their burly, still-green boughs, offer some protection from heavy snowfall as well as a break from the bitter winds of the Catskill winter. Sometimes I wonder how anything survives here, especially when the temperature falls below zero, as it often does now at night. The other morning, the outdoor thermometer visible through the kitchen window touched -20 degrees—which made me look twice, to be sure—a dire reading, the coldest of the season. Even the coyote's howl, that sad lament, which is usually the only sound to pierce the winter night, other than the reckless wind, assailing the windows, has stilled. The coyotes must be suffering too, especially with such a deep snow pack thwarting their nightly travels, and with prey so scarce.

Deer can survive winter better than most species. Their fuzzy coat, grayish now, no longer the rich red-brown of summer, keeps them warm. The wooly underfur is dense, insulating, while the outer layer has coarse hairs that are hollow, trapping air that also aids in insulation. In fall, deer gorge themselves, as many critters do, to prepare for winter. Their diet, rich in nuts and berries, builds fat reserves that help sustain them in leaner days. Their metabolism also slows in winter, which allows them to survive on twigs, stems and other woody browse, and to save energy they don't have to spend on foraging. For deer, conservation is everything; they take not one wasted step. Still, an early winter that lengthens into spring can spell starvation, and fawns especially are vulnerable. Adults increase their body weight by as much as 30 percent but fawns gain only about half that—most of what they eat goes into growing muscle and bone. White-tailed deer are a miracle of adaptation, and their story is about survival of the fittest.

The deer scatter as I open the back door, retreating into the woods. Their sharp hearing, as well as their speed, are their principal defenses. As I head out, determined to make the most of this balmy winter day, finally no longer in the single digits, the squeaking of my snowshoes alert any critters of my approach. Aficionados of wooden snowshoes, the traditionalists of the sport, complain aluminum shoes scare away the wildlife, while theirs glide silently. Still, it is a sacrifice I have been willing to make. Snowshoeing is hard work, and the aluminum shoes are lightweight, as are the poles I use now to propel myself forward.

I have seen only signs of the wind so far, rippling the snow like surf on sand, sculpting dunes of the purest white. Finally, I spot some tracks, the trail of a mammal light enough to traverse the snow's frozen surface. It's an eastern cottontail, judging from the prints' U-shape, revealing how its feet fell—the front, landing first, then the back, landing in front and to the sides of these. I follow the tracks only a little way before they disappear. Then I spot some scraps of fur, grayish-brown, the cottontail's color, as well as a dash of red, blood most probably. And there in the snow is a suspicious scrape, easy to miss, made, I presume, by the edge of a raptor's wing, brushing the surface. The end of this winter's tale was abrupt and violent. The hapless rabbit, perennial prey which finds no haven in any season, has been snatched by a barred owl, resident here year-round, its talons remorseless, unforgiving.

Snowshoeing into the woods, I take a trail I clear every spring of rocks and branches, winter's leavings, which the animals, always opportunistic, also use. In the mud season in the Catskills, which can last into May, even June, I will find many tracks here, including wild turkey and white-tailed deer, but not today—the snow is too deep. At a hemlock tree, I pause to examine some twigs and slender branches, which have been broken off and chewed, lying in the snow. A porcupine could be lurking in the boughs above—the hemlock is a favorite of his, its thick evergreen foliage providing a safe hiding place as well as food—and these might be the remnants of a morning meal. On a hunch, I clear away the snow at the base of the tree and find the bark has been diligently gnawed and also scratched by sharp, long claws, signs of the porcupine's presence as persuasive as the wide, furrowing trail he makes when winding through deep snow.

The wind, which has been bearable, suddenly sharpens—I am beginning to feel the cold. But it's a short way to the black walnut tree so I snowshoe on. In summer, I would seek the cool shade of its luxuriant canopy, returning home from a hike and in fall, I loved how its fernlike leaves turned bright yellow, contrasting with the rich dark brown of its rugged bark. It was struck down by lightning several summers ago (the *kraaack!* reverberating through the forest, so loud I could hear it in the house), its carcass and skeletal branches a marker for me in the winter woods, where everything looks different, swallowed by snow. Ahead I see the jagged trunk of the ruined tree, rising above the snow, a welcome talisman. The tree had been hit fairly high up and another section had broken off later, so that two logs, both generous with branches, now lay on the forest floor. The black walnut was not the tallest tree in the woods, thirty feet high or so, nor was it particularly out in the open. Why light-

ning chose it, I had no idea. But that is why lightning is so terrifying—
it is unpredictable, its ire not always justified.

I intend to turn around as soon as I reach the stump but when I
arrive, something catches my eye. Under the thick tangled branches of
the fallen tree, beneath a heavy log, is what looks like an arc of black, set
in the dazzling snow. My eyes, used to the white, take a few seconds to
adjust, but then I see them: The black bear family, mother and twins,
which I first glimpsed last May, when the cubs were new. This is their
den, I realize, my heart racing. Mama would have made it for them,
digging out a hollow, lining it with leaves and branches of soft pine—
perhaps even using the red wool blanket I'd left out to air, which mys-
teriously disappeared last fall—choosing the black walnut tree, still
sheltering critters as it did in life, as their winter home. The mother is
curled around her twins, sheltering them with her formidable body,
while the cubs, cuddled together, press against her belly, seeking the
warmth of her soft, dense underfur. They will sleep like this, lost in each
other, well into March, even early April, when the long Catskill winter
finally recedes.

I take another few steps to get a closer look, silently cursing my
squeaking snowshoes. Too late. An eye pops open, one of the cubs—it's
hard to tell where one begins and the other ends, they are in such a jumble of glossy black fur. I am breathless when I see the cub is awake, afraid
to move. Black bears are light hibernators, easily awakened, but this
mother, at least, seems slow to rouse so I am not afraid. I am more worried that I will frighten the cubs and interrupt this idyll, disturbing the
solace of their winter sleep. It is time for me to go.

Stepping back, I take one last look and realize, yet again, the perfection of nature—its genius of adaptability, the brilliance of its beauty.
The sight of these bears in their den, the mother embracing her cubs,
asleep under the boughs of a fallen black walnut tree, their jet black fur
brushed by the white of newly fallen snow, has been the gift of this winter—and perhaps of a lifetime.

THE CROWS ARE ALWAYS WATCHING ON THE MOUNTAIN, AND IN WINter especially, they watch for me. The "sentry" at his lonely post atop the tallest aspen tree will spot me leaving the house, trudging through the snow to the feeders, bearing buckets of black oil sunflower seed, and on the coldest days, the crows' favorite treat, peanuts, still in the shell, roasted and unsalted. As I approach, the sentry signals his clan—the "*caw-caw-caw*" that spoils winter's silence, along with the crunch of my boots in the snow, is a communication as effective as Morse code. Within seconds, his cronies are flying in, alighting in the branches of the quaking aspen, waiting for me to depart—ever distrustful of their human benefactor—before seizing their prize.

I can always count on finding crow tracks around the feeders, especially after a snowfall. The crow's foot, sturdy and larger than that of most birds, leaves a distinct print. Three toes point forward, the fourth, just as thick as the others, points back—the entire track is about three inches long. In last night's light, wet snow, which has covered the remnant of the winter snowpack, now turned to mush by the warming temperatures of late March, the claw mark at the end of each toe is visible as well. The manic cross-hatching of turkey tracks is noticeably absent this morning. Yesterday, I scattered just peanuts for the crows (prized by blue jays and of course the squirrels, the sprightly reds stealing them away from the grays, infuriating their larger cousins), which are of no interest to the turkeys. It is the last time I put out peanuts, as the bears will be out of hibernation soon, and the feeders, if I don't want to find them crushed, halfway up the mountain, must come down soon.

Winter has been long in the Catskills, snowy and bitterly cold. But the vernal equinox last week has marked the season as new. With spring have come days which now seem luxuriously long, and even when it's still stubborn with cold, the sunlight, stronger now, has worked to vanquish the snow. The worry, at this time of year in the Great Western Catskills, is flooding—when rains conspire with melting snow to cause rivers, choked with ice, to overflow and swollen streams to break their banks. But this spring, thankfully, the thaw has been gradual, with slowly rising temperatures and gentle rains. Until last night, when several inches of new snow fell, the ground was visible for the first time since December. It is no small thing, to see the earth again after so many months—to see the pledge, the first brave hint, of spring, and to dream of summer here, the growing season, made all the more precious by its brevity.

I have not seen the deer family—the doe and her twins—for well over a week, when I watched them weaving their way through the upper meadow, the snow no longer restricting their slow steps. But I was relieved, early this morning, to spot another doe and her fawn, emerging tentatively out of the woods. I barely glimpsed them this winter, and worried they might not have survived the inexorable cold. At the feeders, where I have found not only tracks but also peanut shells, telling the tale of the crows' visit, I also find the tracks of this doe and her single fawn. They pass by the feeders, heading toward the stone fence that marks the property's western boundary—the rocks jagged and irregular, piled up decades ago by the resolute farmer who worked this land—where the tracks disappear. I assume the deer have jumped over the fence. But on a whim, I decide it would be fun—and a challenge—to backtrack them, to see how far I can follow their trail into the woods.

Along the way, I encounter the inevitable turkey tracks. But this set of tracks is intriguing—on either side, running parallel to them, are straight lines etched in the snow. I am puzzled at first but then realize this trail has been made by the tips of tom turkey's wing feathers. I spied him this morning in full breeding regalia—his vibrant blue head, his fierce red wattles, his brownish feathers iridescent in the sunlight, flashing hues of red, gold, copper, green and bronze—fanning his tail feathers for his harem of five hens. When displaying for the females, the tom shivers his wings, and drags them along the ground, leaving these marks, though in the snow, a sure sign of spring.

I head into the Cut, trailing the doe and her fawn, their tracks still sharp in the recent snowfall. The ground, starting to thaw, and the rocks, now reappearing, without deep snow to seal them, make this a slippery path, and I am glad I have Trusty Tree, my walking stick. I pass a stand of stately sugar maple trees, their dark-brown, furrowed bark an indication of their age, and following the fading tracks, climb over the jumble of bluestone rocks that were once a foundation, probably of a small sugar house. The sap is flowing in the maples now, the sweet, sticky liquid that will become maple syrup, a deliciousness that is unique, unassailable. March is maple sugaring season in the Catskills, perhaps the surest proof of spring here, when the trees are finally freed from their torpor by temperatures rallying past freezing, and the sap begins to rise, their lifeblood returning.

The woods are beginning to close around me, though it's brighter than it would be in summer, as the trees are bare, their branches rattling in the chilly breeze. Then I see something ahead, in the snow. Approaching it slowly, cautiously, ever on the alert, especially since my encounter with the hibernating bear family, I realize what it is . . .

A fawn, lying on its side, brushed by last night's snow, its soft dark eyes open, staring, the light of life extinguished.

I have to sit down and I do, in the snow, next to the dead animal. Is this one of the twins? Is the doe and fawn I saw today, whose trail has led me here, its mother and sibling? This fawn has not been dead for long. There are no signs, yet, of predators—it has not been savaged by coyotes or taken down by bobcats. There are no wounds that I can see. I surmise instead that the fawn has starved to death or perhaps, the weaker twin, it was more susceptible to the harsh elements. What the experts would say, I realize, is that this is the deer's story—that the strong survive and the weak perish, for the greater good of the herd. But that is a cliché that does not console me or still the sorrow I feel at the sight of this fawn, still so new, fuzzy in its gray winter coat, felled in the snow. And it's spring—how did the fawn die in spring, having escaped the worst of winter, with sweet green grass a promise away? It just seems so unfair, though fairness, I realize, has nothing to do with nature.

The doe and her surviving fawn have moved on, and I must, too. But I am reluctant to leave. The fawn, I know, will feed the critters that live on the mountain—the coyotes, the crows, the bobcats, and the turkey vultures . . . The cycle of life and death is natural here. It rules everything, I know that, too. And yet, try as I will to accept it, to find the objectivity of the scientist, the equanimity of the philosopher, sitting beside this beautiful creature, perfect even in death, covered by a light shroud of snow, I am heartbroken. That is my story, my very human story.

CHAPTER NINE

Ursus Americanus: Our American Bear

(For Alf Evers, Homo Catskillicanus)

"H*uff . . . huff . . . HUFF! . . .*"

I looked up from where I was crouched over the unruly sea-green juniper, loppers in hand, on the grassy lawn, newly mown, that is wedged between the upper and lower meadows, which, in late May, were already high with wildflowers and tender green grasses. There, no more than seventy-five feet away, standing on all fours, was a black bear, *Ursus americanus,* described in my *Peterson Field Guide to Mammals of North America* as "the largest living terrestrial carnivore" in North America.

I knew this bear, had seen her two springs before as a yearling (more accurately, at sixteen to seventeen months, as most cubs are born in the Catskills in January or early February)—then, last year, as an adult black bear. But in those encounters, the advantage was to me. That first time I glimpsed her—the first time I ever saw a bear in the wild—she was young, smaller and more easily intimidated, recently expelled by her mother, whose mandate to breed, triggered by estrus, forces her to drive her young away, saving them from courting males, likely to kill any cubs in their path. I was walking up the incline of the dirt driveway in the early afternoon and, out of the corner of my eye, saw a dark figure poking at the sunflower seed feeders I had been foolish enough to leave up in spring. I thought, at first, it was one of the wild turkeys that patrol the mountain, which are tall enough to peck at the hanging feeders. But when I turned, I was startled to see a young black bear, standing upright, both paws on the feeder, nearly three feet tall.

I clapped my hands, then yelled. The bear gamboled away on all fours with the slightly comical, uneven gait of any young animal, pausing where the woods begin to overtake grass. She hid behind a skinny maple sapling and peeked out at me, convinced that she was concealed by the young tree. I started clapping again and she slipped into the newly leafed woods, her disappearance, oddly, filling me with regret. No need. The next day, at high noon on a sunny day, she was back, peering forlornly at where the feeders, which I had finally taken down, had been. This time, though, the bear was reluctant to leave and I had to augment my repertoire of clapping and yelling with that old camping stratagem: pot banging. Finally, the little bear took off, though I saw her later, still somewhat hidden but looking out from a thick stand of willows.

I did not see her again until the following summer when, full grown, she would occasionally visit, mostly at dusk or dawn, still searching for the sunflower seed feeders, which I did not put out again during bear season, saving them for the winter months, January to April, when black bear in the Catskills hibernate. That she was so persistent was not a surprise. Wildlife biologist Larry Bifaro, a bear specialist from the New York State Department of Environmental Conservation (NYS DEC) I had contacted on first sighting this bear, had informed me that black bear were famous for their memory and sense of smell. "She'll remember a feeding station easily for a few years," Bifaro told me, and black bear were also capable of scenting a favorite food "a mile or two away."

Though I no longer put out feeders, I would still leave black oil sunflower seed for the chipmunks, my comical, irresistible companions, who, when I went inside, would sit up, stretching their necks to catch a glimpse of me in the house, chirping insistently for their snack.

On one warm, early summer night, my windows open to hear the *"gunk-gunk-gunk"* of green frogs calling from the pond, I startled awake at the sound of snuffling, so close it sounded as if it was coming from the foot of the bed. When I peeked out the window, just feet away from where I stood, digging at the ground with her claws, rooting through the grass for any stray sunflower seeds the chipmunks had overlooked, was the bear. Before she could catch my scent I shouted "Git! Git! GIT!" I was surprised at how easily I routed her and stunned to see, in the moonlight, how fast she could run, sprinting for the woods.

She and I would repeat this scenario several times that summer—me, shouting, she, running, and I feeling safe and secure behind the "blind" of my bedroom window. There was always a thrill at sighting her and perhaps, I have to admit, a sense of power that I could scatter her so readily. But I had never been caught out in the open with a full-grown black bear before. My yearling was now an adult—an animal that stands two to three feet high at the shoulder and ranges four and a half to five feet long, from snout to stubby tail. Even this bear—if it were the same, probably a female—would average 170 pounds. (According to NYS DEC, a female can weigh in at 150 to 250 pounds, while males top the scales at 200 to 600 pounds.)

"Huff . . . huff . . . huff . . ."

The bear's huffing—a low blowing I'd sensed before I heard, a strange sound that raised the hair on my arms, at the back of my neck—indicated its own agitation, even fear, I would learn. She stood there, on all fours, at the Cut. I have delighted, at dusk especially, in seeing so many critters emerge from the woods there, usually on their way to the pond—porcupine, fox, raccoon, opossum, even a lanky coyote, loping along like a domestic dog returning home, which I thought it was for a second, its gait was so assured. And I knew this was the access for the bear as well—I had seen her as a youngster, then as an adult, on her way back into the woods, fleeing with that surprising speed.

Black bear, I reminded myself, rarely attack people. They are basically shy, reclusive animals that, as I had found, will run rather than risk confrontation. But as I stood there, where I had been pruning dead branches off of the juniper, I had to fight down fear, such a primal sense of danger. The bear was stronger, so much stronger, so much more powerful . . . I felt, for the first time in my life, like prey, vulnerable and trapped.

"Don't run," I told myself, "and don't turn your back on it."

"It," I remember thinking. Before, this bear had always been "she"—but now, as we both stood there, seemingly frozen in our mutual fear, the bear had become "*it*."

IN *THE CATSKILLS,* HIS VIRTUOSIC BIOGRAPHY OF PLACE, ALF EVERS writes "in the infant years of the nineteenth century, the bear was the lord of the wild Catskills." But with the coming of Europeans to the region, the bear's riches of fat, meat and fur made it a ready target. And, as Catskills settlers cleared land for their farms, the bear, an accomplished omnivore, was increasingly in competition for prized native fruits and nuts, such as blueberries, raspberries and huckleberries, sweet chestnuts and black walnuts—and all too often was demonized for raids on cultivated crops, corn especially, and even on livestock such as chickens, rabbits and sheep. But it was avaricious and ruthless commercial interests that would drive the black bear and other large mammals, notably white-tailed deer, from these mountains—and even cause the disappearance, for a time, of trout from once pristine, chilly Catskill streams.

The stripping away of Catskill forests, especially in Ulster County's "High Peaks," of ancient stands of Eastern hemlocks—whose bark contains tannin, a chemical once employed in the tanning process to convert cowhide into that crucial nineteenth-century commodity leather (used in household as well as military items such as shoes and clothing, bridles and saddles)—would be at the epicenter of America's first great environmental catastrophe, and by the 1850s, mark the demise of the Catskills as its first true wilderness. Tens of millions of hemlocks were slaughtered in the eighteenth and nineteenth centuries, leaving only those that were, at the time, inaccessible to the tanner's axe. (Today, some old growth forest can still be found, primarily in the vicinity of Slide, the Catskills' highest peak, and Big Indian mountains—an estimated 53,400–63,000 acres, the largest of which is about twenty-five square miles.) The waste was staggering, and still appalling. The stripped-bare carcasses were left to rot on mountainsides—hemlock, a "softwood," was considered a poor building material, its wood splintering easily. Even more tragic is that only the lower part of the tree's trunk was deemed valuable for its tannin. Hemlocks standing over one hundred feet tall, some whose diameters were easily over four feet, were routinely felled. The white corpses of the hemlocks, left in place to dry in the sun, especially in drought, were subject to frequent fires, and the Catskills' once verdant slopes became black and charred, on certain days, the mountain landscape clogged with thick, acrid smoke. Without the hemlocks and other trees, such as spruce, pine, basswood, poplar, beech and white birch, also harvested for lumber and charcoal (subsequently, even second-growth hardwoods would be exploited to make barrel hoops), the thin rocky soil would erode, washing down the mountain, choking streams that were already polluted by toxic foul-smelling wastes dumped into them by careless tanneries. Forests in the Catskills, featuring conifers such as hemlock and white pine, sturdy trees that can support a bear climbing them, which, due to their furrowed bark, mother bears prefer—easier for a young bear to grip as it scrambles to safety—were decimated.

Alexis de Tocqueville, noting this disregard of the natural world in favor of the exploitation of its resources as early as 1840, wrote in *Democracy in America* (Vol. II, Ch. 17):

> In Europe people talk a great deal of the wilds of America,
> but the Americans themselves never think about them: they
> are insensible to the wonders of inanimate nature, and they
> may be said not to perceive the mighty forests which surround

them till they fall beneath the hatchet. Their eyes are fixed upon another sight: the American people views its own march across these wilds—drying swamps, turning the course of rivers, peopling solitudes, and subduing nature.

(Translated by Henry Reeve.)

In the Great Western Catskills, the Catskills where I live, with their rounded foothills and gentle slopes and broad, sunny valleys, many hillsides had also been cleared for sheep and cow pastureland, as the rocky clay soil, offering "two stones for every dirt," had made agriculture, for commercial purposes, largely too difficult—and unprofitable—a challenge. Logging had taken a toll, too—indeed, some Delaware County residents still claim that the mainsail spar for the USS *Constitution*, "old Ironsides," a wooden-hulled, three-masted heavy frigate commissioned by George Washington in 1797 (the oldest naval vessel still in service) was toppled from these hills. And a trip to legendary Catskills essayist and naturalist John Burroughs' family farm in the Western Catskills town of Roxbury reveals a view of wooded hills that in Burroughs' day, in the later nineteenth century, would have featured cleared hilltops with only scattered clumps of forest, a landscape dotted with stone fences, used as enclosures for livestock, and grazing herds of sheep and cattle.

Burroughs, a prolific writer who popularized nature and the Catskills, arguing for stewardship of the land while warning against the rampant exploitation of resources, would see the Forest Preserve Act enacted into law in 1885 in New York, a law which originally targeted Adirondacks woodlands, to protect them from the degradation that the Catskills had suffered, specifically to protect waterways such as the Erie Canal, linking the waters of Lake Erie in the west and the Hudson River at Albany in the east, and therefore, the state's water-dependent commerce, from the silting that is a by-product of soil erosion. (Ironically, the Catskills were considered too despoiled to be included in this legislation, their rivers and streams, which now fill New York City's upstate reservoir system with the sweet, pure mountain water that supplies millions of downstaters—"flatlanders," in local parlance—said to be insignificant.) Tucked into that bill, however, were 35,000 acres of land around and on Slide Mountain, the Catskills' tallest peak (4,180 feet). Ulster County had persuaded New York State to accept the parcel in lieu of an outstanding $40,000 tax bill. In 1894, a constitutional amendment, Article 14, codified both the Adirondack and Catskill Forest Preserves as "forever wild":

> . . . they shall not be leased, sold or exchanged, or be taken by any corporation, public or private, nor shall the timber be sold, removed or destroyed.

Today's Catskill Park, which is a partnership of public and private lands, totals some 700,000 acres in several upstate New York counties—the Forest Preserve, "Forever Wild," as every sign proclaims, now counts nearly 300,000 (287,000) acres.

In a sense, the eco-holocaust that the Catskills suffered was also a call to conscience that marked the beginning of the modern conservationist movement. These Americans, at least, did *not* fill de Tocqueville's bill. Thomas Cole, founder of the famed Hudson River School, the nation's first indigenous school of art, painted idyllic views of the Catskills' craggy heights and forbidding cloves, popularizing his own luminous vision of a wild, untamed American wilderness, still pristine in the 1820s; Ralph Waldo Emerson espoused, in his landmark work of Transcendentalism, *Nature,* published in 1836, the idea that divinity suffuses all nature, and that through the experience of being one with nature, an inner "transcendence" is possible; Henry David Thoreau, whose classic *Walden or*, *Life in the Woods,* published in 1854, chronicled his experiment in rustic living, was the earliest advocate for setting aside wild lands and leaving them in their natural state, as early as the 1830s; and Walt Whitman, writing in the mid-nineteenth century, who believed that nature was the root of all beauty, celebrated the unique relationship of Americans to their landscape in numerous poems and essays. These men were all, in a very real sense, proto-preservationists. A classic though now little-known work, certainly informed by the Catskills' environmental carnage, was *Man and Nature, Or, Physical Geography as Modified by Man* by George Perkins Marsh, published in 1864, which made the then revelatory connection between deforestation, soil erosion, and desertification—it caused a sensation (a sort of *Silent Spring* of its day) and had a direct influence on the establishment of the Forest Preserve.

With the establishment of the Forest Preserve in 1885, and then, with the building of the first of the great upstate New York City reservoirs in 1912 —the Ashokan ("place of fish" in the Lenape language)— the Catskills had won protections that would allow its mountains, streams and valleys to regenerate. The bark-tanning industry that once saw over sixty tanneries despoiling the Catskills landscape had essentially ceased by the end of the nineteenth century (synthetics would replace both tannin and leather—alas, too late for the Catskills' old-growth hemlocks, those majestic sequoias of the East), and most of the bluestone quarries (that handsome, durable, readily mined rock that a century ago was New York City's predominant sidewalk material) that had scarred mountains and hillsides throughout the Catskills had closed, ceding to cheaper and more readily available Portland cement. As farms were increasingly abandoned in the depressed economic era of the early twentieth century, the land left to nature, the black bear started to make its slow return, along with the forests and the mountains where it had always thrived. New York State began to stock streams with trout—now finally clearing from decades of pollution and shaded by second- and third-growth trees, which cooled the water so this popular game fish could once again thrive. And even white-tailed deer, once widespread, were slowly reintroduced—they were first penned in an enclosure at the foot of Slide Mountain where they became a local attraction, so long had it been since they were seen roaming the Catskills. Now, according to NYS DEC figures, the Adirondack Park, whose forests were spared the savagery inflicted on the Catskills, has 4,000–5,000 black bears, and the Catskill population has rebounded to 1,500 to 2,000 individuals, the highest density population in New York State. The black bear is once again the lord of these wild Catskills.

Even so, it was a surprise, that mild day in January, when I first realized there was bear on the mountain. I was looking through my binoculars toward the western woods where I hang a variety of seed and suet feeders for winter birds. My usual concern is gray squirrels, which, no matter what promises feeder manufacturers make as to their being "squirrel proof," I have never been able to outsmart. That canny, clever, conniving critter—the gray squirrel—will defeat any baffle, chew through the hardest plastic, even, eventually, metal, and will out of spite find a way to render useless any feeder that it *can't* penetrate. When it comes to squirrels, I have to admit, I play Elmer Fudd to their Bugs Bunny.

But this time the feeder, a hefty triple tube job holding ten pounds of black oil sunflower seed, had simply vanished. What had those devilish squirrels done with it? I bundled up and traipsed out to the feeders, the muddy earth slick and soft with January thaw—and was stunned to discover that all that was left were some scattered plastic shards, pieces of the feeder's base. I gathered them up, then noticed for the first time that the branch—a sturdy, thick lower branch that I'd hung the heavy feeder from—had been snapped from the sugar maple tree and was lying on the ground. I held the plastic shards in my hand and thought *"Bear..."* Who else was strong enough to snap that branch, then cart off the triple tube feeder?

Then I noticed even more compelling proof of black bear. By Woodchuck Rock, the worn bluestone boulder under which the resident woodchuck was now hibernating, there had been, the previous fall, a yellow jacket nest. Yellow jackets are vicious—not a word I enjoy applying to any critter, but in this case, true. These predatory wasps often make nests in the ground or in hollow tree trunks, and they have been known, the colony acting together, to sting people who happen on them to death. A contractor friend once told me about stumbling on a nest while working on a house, how he ran from the swarm, just reaching his truck in time. He claimed that the yellow jackets pursued him furiously for a mile before finally giving up their chase. ("And that," he concluded, "ain't no Catskill tall tale!") I had discovered this particular nest late last summer when I wandered out to the feeder area, already thinking of my winter birds. Right by Woodchuck Rock, I saw it—a hole in the ground the size of a quarter, topping a slightly raised mound of earth, and as I approached, a solitary yellow jacket flew out and hovered there, watching me warily—a sentry for the colony. It concerned me—not for winter, as I knew the cold would kill off most of the insects, but for the following spring, when the queen would emerge and establish a new nest, perhaps close by. A yellow jacket nest so near a residence is one natural thing that can't be tolerated—these wasps are too dangerous, with their ability to sting repeatedly, roused to anger even by the hum of a power mower.

But now, as I turned to go back to the house, plastic pieces in hand, I noticed in the icy mud of this mild January day, what seemed to be a huge gouge in the ground. The gouge, I realized, was precisely where the yellow jacket nest had been—or the entrance to it—that raised hole in the ground now no longer identifiable. Then I saw the ragged claw marks, clear in the snow melt and mud, at the edges of this excavation. The bear (chances are it was a "he," as females, pregnant or denned up with young, usually won't venture out until spring) had snatched my very pricey triple-tube bird feeder, helping himself happily to his prize of black-oil sunflower seed, his January thaw repast. But in hopes of spiking his diet with protein (bears covet not just honey but honeybees too, and also relish wasps, seemingly oblivious to their stings), the bear had sniffed out the yellow jacket nest, then scooped out what was left of it.

I was amazed at the bear's strength—he had snapped a sugar maple branch, solid and durable, casually carried off an unwieldy, weighty feeder that I had struggled just to lift, and then plundered the still-frozen ground in search of the yellow jackets. I had never encountered such strength in nature, such easy power. And as I hurried back to the house I looked over my shoulder, for the first time truly aware of the wildness of this place.

I COULDN'T TAKE MY EYES OFF OF THE BEAR AS IT STOOD, ON ALL FOURS, at the entrance to the woods. The bear, I noticed, would not look directly at me. It moved its head from side to side, as if it were deliberately refusing to make eye contact but was watching me with its peripheral vision. Making eye contact can be interpreted as issuing a challenge, a prelude to aggression, which the bear, wiser than I, wanted to avoid.

The bear, except for its occasional "huffs," was quiet. Used to depictions of bears as ferocious, growling beasts, I would be surprised to learn that black bears are largely silent critters, though they can communicate through a variety of sounds, such as grunts, teeth clacking (indicating fear, as does the "huffing" I had witnessed), moans, and tongue clicking (a common way for mother bears to signal their cubs). But bears *don't* growl in the literal sense; North American Bear Center researchers report that in "over 40 years of capturing, observing, and confronting bears—and seeing the occasional fight—they have yet to hear their first growl." Another tidbit: that growling bear of movie and TV fame is most likely a wolf doing a voice-over.

The fierce, growling black bear is such a staple of American folklore, as is the brave hunter who slays it. Davey Crockett, the legendary "King of the Wild Frontier," was so heroic he "killed him a bear when he was only three," as the popular ballad goes. Perhaps one of the bear's most iconic depictions—a taxidermy specimen rearing up on its hind legs, arms outstretched, claws flashing, its mouth contorted in a snarl, caught in that moment just before it charges at, we presume, the hunter who felled it—is also apocryphal. When a bear stands on its hind legs, it is not assuming an aggressive stance. Bears stand out of curiosity, to gather information, to assess a situation. Chances are, the perceived dangerous, predatory bear, now frozen forever in a stance it never assumed in life, was simply ambling through the fall woods, looking to gorge on berries and acorn mast, as it went about its foraging to fatten itself for hibernation, when the hunter shot it.

Finally, hoping to end the standoff, I took off my wide-brimmed canvas "courage" hat that had accompanied me rafting on the Colorado River, through all 277 miles of the Grand Canyon, still and forever stained red with river mud, and started to wave it at the bear. "Git, Bear, git! GIT!," I yelled, jumping up and down, arms flailing, trying to make myself as imposing as possible.

Abruptly the bear bounded away, heading for the upper meadow, where she ran through the May grass, already high enough to hide her from view. She emerged from the meadow at the pond, then paused, looking back at me before crashing into the woods. She was, I suspected, on her way to the clearing where I hung the winter feeders. Just a few weeks before, when the trees were not yet fully leafed, I'd glanced out a kitchen window and caught sight of a dark shape shifting through the woods in the early morning sunlight. It crossed my mind that it might be the bear with her stubborn memory, checking to see if the feeders were still up (which they weren't), hunting for stray sunflower seeds or a scrap of suet the squirrels had neglected.

After several minutes, as I watched, transfixed, the bear retraced her route through the upper meadow, the tall grass rippling in her wake, tamping down a path that I would see her, as well as white-tailed deer, follow that summer. I expected her to head back into the woods, but she kept ambling in my direction. She stopped at the mown path, adjacent to the lower meadow, which leads down the mountain, and this time stopped no more than twenty-five feet from where I was standing. She

stood there in silence, shaking her head from side to side, not looking directly at me but watching me. Bears are among the most intelligent mammals, with a large brain compared to their body size. And they are, as is typical of intelligence, very curious critters. She was certainly curious about me as I was about her—two mammals, of different species, one on two feet, the other on all fours . . .

Her jet-black fur was thick and lustrous, the brown of her muzzle—her only other color—made her look less fierce, endearing even. She showed no sign of aggression or fear—she wasn't huffing or clacking her teeth. She knew me, this bear, had seen me first as a youngster two years before. I had never harmed her or chased after her, she knew that too. But bears are still wild animals and they are unpredictable, which is another trait they have in common with us, along with curiosity and, of course, intelligence. And for her own protection, above all, I could not allow her to become accustomed to being in people's backyards, especially in broad daylight. Suddenly I began jumping up and down, swatting at the air with my river hat, yelling at the bear to "Git!" Finally, she did, galloping, cutting across the mown path and disappearing into the woods. I could hear the crackling of fallen tree branches and last year's leaves as she scrambled through the understory, then silence. I waited a few minutes, still calling "Git, bear, git!" Then I walked slowly, not wanting to run, turned sideways so that my back wasn't completely exposed, back to the house.

THE FIRST SPRING I WAS IN THE CATSKILLS I RESTORED A BOGGY AREA in the woods that had, as its source, the same spring that provides the mountain water for my home. This bog, though overgrown with leafy greens that leech water out of the moist soil and embedded in a small forest of wild rose bushes, I knew, if cleared, could be a running stream—a benefit for critters as well as a lovely place for me to pause in my walks in the woods. It turned out to be a laborious task—more than I bargained for—which I accomplished with, among other tools, a machete, cutting down an endless tangle of wild rose bushes, planted in the Catskills as ornamentals but that now have invaded these foothills, crowding out native species. The wild rose succeeds because it is extraordinarily resilient—a stubborn, tough shrub that must be uprooted from the rocky soil to be vanquished. That summer, even after that first arduous clearing, I had to return to the bog to cut back the thorny, spidery branches of the wild rose yet again. Indeed, if I don't intervene every spring, Nature will quickly overtake my efforts, turning my precious small stream once again into a muddy bog thick with weeds and the treacherous wild rose.

What startled me most out of this experience was how angry the wild rose made me—how easy it was to demonize it, to attribute to its prickly ways wickedness and willfulness, as if it were defying me, even trying to injure me. (I have learned, the hard way, to wear protective plastic eyeglasses when hacking away at these bushes—and inevitably, even if I do cover myself, the sharp thorns of the wild rose will find a way to assault me, inflicting small, painful wounds that can easily become infected.) My yearly battles with the wild rose stand in my mind as representative of how Nature, if given the chance, will not just rebound but overwhelm, and this small, exceedingly small, experience gave me insight into the lives of those early Catskill settlers who would penetrate the woods and attempt to carve out a home in the wilderness, where even the earth seems resentful, the rocky soil resistant to planting, guarding itself with "head rocks," large chunks of bluestone quickly encountered when digging in these hills.

As humans, we do have reason to distrust Nature, not only because it is so powerful and utterly indifferent to us, but because in it, alone, vulnerable, with only our brain to aid us, we are prey—our size and strength, our senses—sight, hearing, smell—so inferior to those of other animals. In the Catskills of the nineteenth century, settlers also had predators, gray wolf and mountain lion to contend with. These, along with elk, would be hunted to extinction and would not return with the black bear and white-tailed deer when the mountains were once again forested. The black bear was hunted, too, as a competitor for various foods, and no doubt out of fear—it can cut an intimidating figure, an animal second in size in North America only to the moose. But the bear stands apart from these other carnivores in Catskill lore—tales of the black bear as directing huckleberry pickers to the best patches and helping hunters find their way home lend credence to the idea that settlers regarded the bear as essentially pacific.

Certainly they saw the bear as a useful animal—its thick fur shielded them from the cold and wind, its rich meat sustained them through those long, bitter Catskill winters, and its fat, a minor industry in itself, greased the hair of nineteenth-century European dandies, the Brylcreem of its day. But I also wondered if such seeming tall tales indicated—at least among the early settlers who lived so close to Nature, who depended on its bounty and understood its unpredictability—some unique connection that they felt with the black bear, a critter as wild as the mountains they shared, with such formidable strength and speed, but also an animal that was curious, intelligent, and, as I had found, somewhat shy . . .

The morning following my encounter with the bear, a bright, warm day in May, I headed back into the woods to finish clearing the stream. I had already hacked away the wild rose that had risen insistently, even this early in the season. But I still had to dig out the stream channel, which was clogged with last year's leaves, already turning to muck, as well as to rake away the rocks that the newly rushing waters, released from winter ice, had pushed down into the stream bed. As I entered the woods, armed with a shovel and rake, I felt not fear, exactly, but apprehension. It was an awareness, really, that I *was* sharing this land with wild critters, in a very real, not just fanciful, idealized way. I was entering the bear's territory—certainly it was hers as much as mine and I had to respect that. Not wanting to surprise her, I called out "Bear, bear, bear," while looking up, rather dramatically, into the crowns of the trees as I went. At the stream, just as I dropped the rake and shovel on the ground, I heard a loud snort, which made my heart leap, as a doe, startled by my presence, fled deeper into the forest, turning back to look at me reproachfully. My racing pulse told me that perhaps I was more on edge than I'd admitted to myself.

As I inspected the stream to see what further work needed to be done, I found large fresh tracks along the edge, imprinted in the soft mud. Bears walk on the soles of their feet, as we do, with the result that their tracks are not always clear, unless made in mud or wet snow. Like us, bears also have five toes (their non-retractable claws often do not register), and these tracks, especially of the back feet, elongated like ours, with the heel showing, struck me as so human-looking. I also found, farther downstream, a place where the stream bank had been caved in. I surmised that the bear had crossed there, in a great rush, as she scrambled her way back up the mountain. There was something about her flight that touched me, made me sad, even. I so wanted to see the bear again, but I wondered if I hadn't frightened her away and to be frank, I half hoped that I had.

THE NEXT EVENING, AS DEEP DUSK WAS SETTLING OVER THE MOUN-
tain after what had been a sunny May day, so warm that it held within
it the promise of summer, I was standing at my bedroom window, which
commands a view of the upper meadow and of that entrance into the
woods that is such a ready transit for all the critters on the mountain,
when I saw, once again, the bear. She emerged out of the woods and
stood there, on all fours, as she had when we had encountered each other
the other day. But this time, standing beside here, also on all fours, were
two black bear cubs.

I can't deny, as I write this, how moved I was by this sight. The
two cubs were young, very young—chances are, since this was May
and cubs are born in January, at the latest February, in the Catskills,
during the mother's winter hibernation, that these "twins" were, at
the most, four months old. She paused there, where the forest opens
out onto the mown land, the people's territory of groomed landscape,
though nothing here can really arrest the growth of anything for
long, without utter diligence. The two cubs, I noticed (I think I was
holding my breath—though at some distance, I didn't want to make
the slightest sound) mimicked her every move; when she stopped, so
did they. The three of them seemed to be waiting. The mother bear,
I could tell, was sniffing—for me, perhaps? I was glad I had shut the
windows earlier, when a thunderstorm had rolled through, and
hadn't yet opened them again. Satisfied that it was safe to pass, she led
them out into the open, then turned to retrace the same route she had
taken when I had seen her swing up into the upper meadow, on her
way to the feeder site. Was that where she was going again? I strained,
in the gathering dusk, to see them—but the grass was already so high
that the two cubs, and then the bear herself, soon melted into the
meadow.

Cubs, I remember thinking, *we have cubs . . .*

When I told friends about first meeting the bear, I emailed a photo of one, featuring a bear's jet-black head and brown muzzle, a full-grown handsome fellow not, I thought, looking particularly fearsome. But one of my friends shot back that she was "frightened" by the picture.

"It looks like a wild animal," Leora wrote. "I realize now I've thought of bears as cute and cuddly. But that bear looks dangerous." Her assessment is at the crux of how many of us see the bear, as either beast, wild and threatening, or as plush, stuffed animal, often the first that children know, the teddy bear. Perhaps no other wild animal stirs such contradictory emotions in us—of affiliation and affection, of apprehension and even fear—as the bear.

That childhood bear, it should be noted, drew its original inspiration from the black bear. And that bear, *ursus Americanus,* was deemed a creation of the Great Spirit by some Native Americans, while the grizzly bear was attributed to the Evil Spirit. It is also the black bear that Alf Evers cites as being a friend of the Catskill people and he relates how a "bear cult" once "girdled the northern world and had participants in Europe, Asia and America." Indeed, the "oldest European constructions that may be described as altars were built perhaps seventy-five thousand years ago by bear cultists . . . in Switzerland." The purpose of these, probably, was to appease the spirits of black bear that humans killed for meat and for clothing—the thick furry pelt of the black bear would have been a bulwark against northern winters for those mountain people too. The black bear, in fact, when sharing the same range with the larger, more aggressive, always dangerous brown "grizzly" bear, is the sometime prey of that formidable carnivore. Indeed, the black bear, as opposed to the grizzly, is "conflict adverse." The black bear once shared North America with larger, long-extinct carnivores, such as saber-toothed cats and dire wolves. Black bears that learned to avoid these predators—by hiding and running away—survived to reproduce. Their descendants inherited these non-confrontational genes, with the result that the black bear is essentially a shy, even reclusive critter, especially when it comes to humans. Indeed, according to the Humane Society of the United States (HSUS), black bear attacks on people average fewer than two per year—more people are killed annually by spiders, dogs and lightning. Or, in the words of Andrew Page, Senior Director of HSUS's Wildlife Abuse Campaign, "More people are killed in vending machine accidents."

(On the contrary, the HSUS reports, more than 33,000 black bears are killed every year by hunters—a figure that doesn't even take into account accidental bear deaths at the hands of humans, such as car strikes.)

All that said, the black bear is indeed a "wild animal," an impressively large one, much stronger and swifter than humans. You don't challenge a bear, and you don't attempt to corner it (that would provoke even this relatively laid-back critter to charge). What the bear is, is unpredictable as Nature itself is—you have to respect that, as part of *its* nature. No, the black bear isn't a teddy bear—even though it was a black bear that gave rise to the creation of that child's cuddly companion.

A key to how we see the bear—both our anthropomorphic images of the black bear as well as our depictions of it as a raging beast—is that the bear can rise up and walk on two feet. That standing bear, interpreted in taxidermy and on pulp magazine covers as a charging, menacing creature, as we have seen, is a false rendering. The bear stands to see, to survey its surroundings, to gather information (and even food). But I can honestly say, the first time I saw a black bear stretching to its full height (I guessed about six feet, though its upraised arms made it appear significantly taller), carelessly stripping fruit from the branches of a black cherry tree at the edge of the woods, I was intimidated. It was easy for me to imagine the fear Catskill settlers might have felt at such a sight—at how overwhelming the black bear, as adversary, would have seemed. But as they came to know this creature, some, at least, saw it as a friend that walked these mountains much as they did—on feet with five toes that rest on the ground as people's do. And this ability of the bear to stand up erect and walk, a trait that is shared only with primates and specifically, us, is certainly an aspect of our identification with the bear. The bear, in a sense, can be seen as a sort of transitional figure between us and nature—it is wild, yet humanlike too, in its curiosity, intelligence and occasional upright bearing, and a bear's carcass offers a veritable bounty. For those early settlers who ate the bear's meat and covered themselves in its fur, who rendered its fat into oil to cook and to light their lamps in the enveloping darkness of Catskill nights, they must have felt very close to this critter indeed.

W E KNOW IT AS "GROUNDHOG DAY," THAT DAY THAT TIES US TO THE animal world, from which we, now living in cities and away from wild nature, are increasingly apart. Though with no predictive (in any scientific sense) value, this rite of rousting a hibernating rodent from its den on February 2nd to see if it sees its shadow (which supposedly predicts six more weeks of winter), is a much-anticipated annual event. And even those of us who pooh-pooh Groundhog Day can't resist sneaking a peek to see if Punxsutawney Phil, or one of his many local furry protégés, rudely awakened and emerging bleary-eyed from his home, spies his shadow or *not*—in which case, there will be that hoped-for early spring.

There are many legends associated with this custom. But well into the twentieth century, this day in the Catskills was known as "Bear's Day." Nineteenth-century Catskill resident and mountain rover, the Reverend Charles Rockwell, wrote that on February 2nd:

> . . . if the sky is clear, the sun shining so that they can see their shadows, and the weather cold . . . they [black bear] sleep quietly on until about the first of April. . . . if the weather is mild and cloudy, they look for an early spring, and often leave their dens.
>
> (*The Catskill Mountains,* 1867)

As the Catskill wilderness receded in the nineteenth century and the mountains were stripped of their majestic forests, the black bear disappeared as well. The groundhog, or woodchuck, which prefers sunny meadows to the cover of dark, shimmering pines and steep rocky crags, the black bear's mountain haunts, eventually ascended to the honor of telling us if there would be an early spring or winter for six more weeks and put its own imprimatur on "Groundhog Day."

"Bear's Day," which, regretfully, has not been reinstated in the Catskills, even with the resurgence of the black bear population (though it would be delightful to see politicians and other types in top hats waking a bear from its torpor to see if it would see its shadow), epitomized this sense, not only of the bear as an intermediary between humans and the mysteries of nature but also of the bear as our friend. The value of "Groundhog Day"—which perhaps for many today has more meaning as the title of a hilarious movie with Bill Murray—has largely been lost on us moderns, tucked away in our warm homes in winter, searching the web for weather forecasts. But for those early settlers scattered throughout the mountains and valleys of the Catskills, struggling through those long, deep, ferocious winters, what could be more wonderful, and ultimately useful, than knowing when winter would end?

On February 2nd "bears of the Catskills were believed to wake from their spell of hibernation and perform certain ritual actions which European bears had been believed to perform many centuries earlier." What I find significant, and curious, is the phrase "their spell of hibernation." Again, we moderns understand hibernation, scientifically, as a way numerous critters, not just some mammals, pass the winter through extraordinary adaptations that allow them to "sleep" portions of the coldest months away. But consider how early humans, including the first denizens of the Catskills, who were, after all, no strangers to spells (just ask Rip Van Winkle, whose own "hibernation" in the Catskill hills lasted some twenty years), might have regarded hibernation. "Spell" seems to fit, rendering the bear, the largest and most noticeable hibernator (indeed, they are just as apt to hibernate in plain view, under a fallen tree, or even a pile of leaves as in a cave, and in the Catskills, a "cave" usually designates a rocky ledge or overhang that is otherwise exposed to the elements), magical, somehow wise, even a wizard of sorts, certainly capable of performing those "ritual acts" that might lead to fathoming even our unpredictable and changeable Catskill weather.

In many Native American legends, the black bear was also profoundly important—as friend, even brother, with whom these first Americans shared the land. For them, the bear's hibernation was linked to the cycle of life itself—the bear "dies" in winter to be reborn in spring, awakening the sleeping earth and restoring life to the frozen landscape. This belief in the bear as the bringer of new life was perhaps heightened by the fact that the bear sow gives birth to her cubs during hibernation, in January and February, the deepest months of winter. It must have seemed a miracle to these human bear brethren to see the black bear emerging from the woods, in April, or late in March, when the snow was finally melting, in the company of her spirited young cubs.

SEVERAL DAYS AFTER FIRST SIGHTING THE MOTHER BEAR WITH HER cubs, I was deadheading the cascading petunia, my favorite flower, its deep purple blooms so fragrant, which filled the flower box of the small wooden shed. That's when I heard it again, that insistent, insinuating sound . . .

"*Huff . . . huff . . . HUFF . . .*"

When I looked up, the bear was there as I had seen her the other day, standing on all fours, where the forest meets the clearing. With her were her twins, and both cubs were standing up, looking straight at me, the family now no more than fifty feet away. I wanted to look at the three of them forever. I could not believe that they were there, so close. The little ones, standing, trying to get a better look at me, didn't come up to their mother's shoulder but they, like their mother, were so curious.

Seeing the mother bear, remembering her as a youngster, now with her own brood, I felt such a sense of protectiveness. A ferocity, an urgency, arose in me—perhaps it was my own maternal instincts. I had known this bear since she herself was a yearling, forced into adulthood by the mandate of her mother's estrus. She had managed to make this mountain we shared her own, and had given me the rare gift of a sense of connection with a truly wild creature.

This bear was, in bear terms, a relatively mild, even fearful critter—running readily when I surprised her, shouting from my window—and now, as she had before, huffing to show her anxiety, her own agitation, but not clacking her teeth at me, or false-charging me, which black bear, certainly with young, can do when they feel threatened. She seemed, frankly, flummoxed.

This time, instead of shouting, I spoke. I spoke loudly but calmly, I did not wave my arms at the little family or my river hat. I stood still and said:

"Bear, turn around and go. Take your babies back into the woods. Hide from me—hide from anything that looks like me. If you ever, ever see me again—*run*!"

The bear watched me in silence, listening carefully, or so it seemed. Then, after a moment, she turned around and hurried back into the woods, her obedient cubs gamboling after her. I stood there for some time, not moving. It was uncanny. Had she understood? Of course not, what was I thinking? But in that moment, the bear and I out in the open, sharing the same ground, without either of us showing aggression or hostility, it was as if she'd heard me, as if we had communicated.

How many times have you heard this: "She's like a mother bear with her cubs . . ."

The protectiveness of the mother bear is storied—and black bears are, in fact, excellent mothers. That said, when it comes to protectiveness, according to wildlife biologist Lynn Rogers, "Sows woof and threaten, sometimes charge to within twenty feet, but generally move off and wait, rejoining their cubs shortly after." Rogers describes the danger of surprising a black bear with cubs as "way overblown." It is worth remembering that as far back as prehistoric days, the black bear's antecedents survived by running and hiding from larger and more aggressive predators—those genes, passed down, are in evidence today in the black bear's shy and typically nonaggressive nature. Perhaps that saying should be amended to "She's like a mother *grizzly* bear with her cubs . . ." Grizzlies, always dangerous to humans as well as black bears, are known to be especially so when protecting their young. Even so, despite these assurances, I would never advance on a mother black bear with her cubs—they are individuals with their own quirks of personality and can be unpredictable, especially when threatened.

The "excellence" of the black bear mother is her devotion and the intimacy of the bond that she shares with her cubs, as many as five every two years (six is not unheard of) but more often two or three. From the time the cubs are born, usually in the Catskills in January (hence the old Indian name "Bear Moon" for that month's full moon), during what we refer to as hibernation but which is really a sort of torpor that essentially allows the bear to sleep through winter without the need to urinate, defecate, drink or eat (as other "true hibernators," such as chipmunks, woodchucks and bats, whose body temperature is more depressed, need to wake up every few days to do), through the following winter, when the cubs also den up with their mother, they lead an almost idyllic existence, in which their mother is their protector, teacher, provider, playmate, and all-encompassing caregiver. There is much for a young black bear to learn, but even with such a vigilant mother, as NYS DEC wildlife biologist Larry Bifaro informed me, the mortality rate for cubs is high—nearly 50 percent perish in their first year, from predation or even falls from a tree. *That* is an essential skill the mother bear teaches her young, who are excellent climbers instinctively. The mother bear knows, however, that all trees for climbing by little bears are not created equal, and her trees of choice are those with width as well as height, and strongly furrowed, such as hemlock and white pine in the Catskills, to enable her young's rapid and sure ascent. Climbing, swimming—these are activities at which young bears, like most country kids, excel. And what perhaps distinguishes black bears, along with their curiosity, is their sense of play.

The black bear, like humans and other primates, and also cats and dogs, all critters that learn and don't just rely on instinct, seem to value play for its own sake—just for the fun of it. The purpose of "play," whether for species bonding or practicing hunting skills, has many theories. But play for its own sake has been demonstrated in bear behavior. Researchers at the North American Bear Center have observed black bears jumping on tamarisk saplings—they climb up, ride the tops down, then climb back up the springy young trees and ride them down again. As they conclude: "Bears eat nothing from these trees, they just seem to enjoy them." And it is certainly fair to correlate this sense of fun and enjoyment, as well as the bear's curiosity, with its intelligence, shown by its brain casing, which, as with chimpanzees, is large in proportion to its body.

Perhaps the most difficult thing for us, as humans, observing this pastoral existence of mother and cubs—this solidity of the loving, even rollicking black bear family—is to comprehend *how*, when they reach seventeen to eighteen months, the cubs, without warning, are suddenly expelled from this familial paradise. The mother may simply abandon them or forcibly drive them away—even, if they pursue her, attacking them. The mother bear, coming into estrus every two years, must drive her cubs away in order to breed again. The family "breakup" is no doubt traumatic for the youngsters, who have been observed crying for their mother. Eventually, siblings will separate too, and if they survive this rite of passage, learning to fend for themselves as they approach their second year, the mother may allow the females a parcel of her own territory— males, when they are of breeding age, will seek their own larger breeding territories that overlap those of several females.

That sharp, sudden, even shocking moment—that moment, dictated by biology, when the mother bear is forced by the mandate of survival of species to abandon her young . . . It is a moment that is so difficult to grasp because the black bear is not only intelligent, curious, playful and affectionate with its offspring, and certainly protective, but because it is a sensitive beast as well, capable of communicating with complex, almost human-sounding vocalizations. But perhaps the black bear mother's "tough love"—so painful for us to contemplate, that image of abandoned cubs wailing for their mother, stranded by her in a tree—is the *ultimate* act of love, to save them from a male black bear, known to kill and eat cubs that wander away from the family group and who will kill yearlings to eliminate them from their mother's care to encourage her to breed again. Ultimately, of course, the little bears, having learned all they can from their careful mother, must shift for themselves to survive. Their rejection by their mother is tough love in the toughest sense of the word—but in the Catskills, and wherever else black bears roam in the wild, that just may be the greatest love of all.

THAT SEPTEMBER I WORRIED OVER THE MOTHER BEAR AND HER CUBS —where were they?—especially as hunting season loomed in late November. Then in October, a friend, whose house is within Mama's range, perhaps five miles or so over the mountain, reported that he saw a "fat" black bear and her two "roly-poly" cubs, scouring the countryside for food as they prepared for hibernation.

That news assuaged me. Chances are, those were my bears, gorging themselves on the bounty of these hills—an abundance of berries and nuts, acorns, fruits such as cherries, grapes and apples and also protein-rich insects and grubs, all fuel for the bears' "big sleep" that can last into April in the Catskills. I did see a sign of bear that autumn—one that taught me a lesson I shouldn't have had to learn.

The nesting box at the eastern end of the upper meadow, long since abandoned by the tree swallows that had claimed it that spring, was hanging at roughly a 90-degree angle off of its four-by-four pine post. It was an Indian summer day in the Catskills, the temperature in the seventies after the season's first frost, the sun warm, the light like burnished gold almost palpable, thick, smoky even. I was inspecting my fall mums—their colors as vivid as the changing foliage on the mountain, the maples, as always, leading the way with reds, oranges, and yellows—which I had assembled on the back deck steps, interspersed with the pumpkins that I would eventually put out for the critters to eat, when I noticed the nest box was askew.

Usually, I would don my sturdy canvas river hat, with its wide, protective brim, leather gloves, and a long-sleeved denim shirt to walk into the meadow to investigate. But I was beguiled by the day, so temperate for October, windless, with no intimation of the harsh winter to come. So I walked up to the meadow and reached for the wooden box, which I also saw had its top torn off, to right it on its pole. That's when a platoon of paper wasps flew out of the box and attempted to assault me. I ran back to the house—paper wasps, slower to anger than yellow jackets, will still sting repeatedly in defense of their nest or when attacked, but at least they didn't pursue me. I realized then that the mangled nest box was the bear's work—whether my bear or another, passing through in search of a spicy snack, paper wasps à la carte, it had clawed out the nest and devoured it, leaving this remnant force, which perhaps had been out on patrol, angry and confused, to defend what was left of their home. Nature, no matter how perfect and pretty it may seem, as soft and consoling as an Indian summer day, always holds surprises and secrets, and sometimes, dangers, too.

It was the following May, almost three years since I first encountered Mama Bear as a yearling, that when I went out, in the early morning, to put my garbage into the can (which I had always been able to leave under the car port, right off the back deck, so polite and considerate was my bear), I saw the plastic can lying on its side. It also appeared that something had sat on it. On closer inspection, I saw that whatever had been in the squashed can was now gone and I headed for the woods, knowing instinctively two things—that this was the work of a bear and that the bear had taken its prize into the woods to examine its haul of, among other things, kitty litter and coffee grounds.

I didn't have to walk far to see, off of the main path, a trail of re-jected human detritus strewn over the bluestone rocks and through the underbrush. Swearing lightly, mostly at myself, I cleaned everything up, while noting that the bear had made off with the recycling—as clean as you can make cat food cans, there will always be a redolence of its con-tents, and clearly this very environmentally correct bear was recycling them for itself. (And how, I now wondered, had that garbage can full of goodies managed to escape not just bears but raccoons too for so long?)

Then, several days later, I spied a small black bear reaching up for the bird feeders in the backyard. She—I decided she was a she, since only a female would be allowed in her mother's territory—stood up to her full height of nearly three feet, and started to pluck at the feeders, full of thistle seed, that staple of the goldfinches' diet, necessary to feed their nestlings, which purple and house finches also enjoy. The other dis-tinguishing feature of thistle seed is that for most critters it seems to taste terrible—not even the ravenous gray squirrels will eat it and after the first few forays, they let the feeders be, leaving the seed spilled on the ground. In the three years that Mama shared this land with me, she never once touched the thistle seed feeders—indeed, she was a *most* considerate and discreet bear, as well as a bear of discerning taste.

But this little one had already taken down one of the feeders, pulled off the top, and was now eating out of it eagerly. My delight in seeing the bear—exactly where I had seen her mother that first time, doing the same thing—was tempered by my concern for how bold she was. She had already stolen my garbage and now she was raiding the feeders in broad daylight (I would learn that she'd also attempted to breech my neighbor's honeybee hives). Both, obviously, were my fault, and I felt bad that I had contributed to her delinquency. I had started to store my garbage in the secure pole barn, putting it out for pickup on Monday morning. The thistle seed feeders would also have to be taken in at night, I realized, and at least for the next several days, as once they have found a food source, bears are faithful in their pursuit of it. I stepped out onto the back deck and clapped my hands but the bear didn't miss a beat (or a thistle seed), and kept on eating, oblivious to my protests. Now I was worried. Here, "nuisance bears"—bears that have lost their fear of humans, that raid garbage cans, feeders, beehives, and especially, those that progress to home invasion, are sometimes killed. So I yelled loudly, clapping, waving my hat, then blowing a whistle for emphasis, and finally, reluctantly, the little bear sauntered off and disappeared into the Cut, where I had last seen her with her mother and sibling the previous May.

I didn't see her again that summer, which relieved me but made me wonder, too. Had she gotten into even more mischief somewhere else? Then, the following spring, early on a May morning, looking out at the upper meadow, I saw a bear enter the high grass from the surrounding woods. But this was no longer Baby Bear—*this* bear was all grown up, her glossy black fur shining in the sunlight. I stood enthralled, watching her go, as her mother had, heading for the pond to drink or perhaps to fish out one of the largemouth bass that populate it, terrorizing frogs and even dragonflies, jumping out of the still waters to snare them. Then, before she reached the pond, she paused and abruptly sat down in the high grass, surveying her surroundings, seeming to sniff the breeze and just enjoying the freshness of a cool morning late in May. She was so comfortable here, she clearly felt so safe . . .

This is her mountain, this black bear, lord of the wild Catskills, and I have the privilege of sharing it with her.

Notes

INTRODUCTION

pg. 14   "Beautiful swimmers":
The Atlantic blue crab's scientific name, *Callinectes sapides*, is derived from the Greek, *kall*—"beautiful," *nectes*—"swimmers," and in tribute to its other virtue, *sapidus*—"savory."

pg. 16   "Age of Fishes":
Middle to Upper Devonian Period, c. 345-370 million years ago.

pg. 17   "Paleozoic Era":
252-542 million years ago.

CHAPTER ONE

pg. 25   "an explanation in science":
Wet snow has five times the weight of dry snow and can absorb more red and yellow light, with the result that, depending on the snow's depth, it can appear, through some icy alchemy, blue.

pg. 26   "their opposites in size":
Black-capped chickadees average five to five-and-a-half inches in length, whereas American crows are about sixteen to twenty-one inches from bill to tail.

pg. 26   "their larger cousins the ravens":

The common raven measures twenty-two to twenty-seven inches in length, comparable to the size of a red-tailed hawk, with a heavy, curved bill (the crow's is thinner and straight). Averaging about forty ounces, a raven is twice the weight of a crow. I have recently begun to see ravens here (usually being chased by crows!), in this part of the Catskill foothills, though naturalist and local wildlife writer Jack McShane reports they are now fairly common in the Andes area of the Western Catskills, which has a higher elevation. They also are at home in the High Peaks Catskills to the East, along with the taller- still Adirondacks. (The Catskill "High Peaks" feature thirty-five mountains above 3,500 feet, with the tallest peak, Slide, at 4,180 feet. The Adirondack "High Peaks," in contrast, includes over forty peaks above 4,000 feet, the tallest, Mt. Marcy, home of the Hudson River headwaters, skies out at 5,344 feet.)

pg. 26   "also a skill of crows":

Even more significant, Goodall observed the Gombe chimpanzees *making* tools by stripping leaves off of the twigs and then using them as implements. (When Goodall's mentor, legendary paleoanthropologist and archaeologist Louis Leakey—whose discoveries, along with his wife Mary, of stone tools and then fossils of early humans in Olduvai Gorge in northern Tanzania in the mid-twentieth century demonstrated that Africa was where our evolution began and where much of it took place—received Jane's news, he now famously declared: "Now we must redefine tool, redefine Man, or accept chimpanzees as humans.") But the crafty crows of the Pacific island of New Caledonia not only use worked leaves, twigs and even their own feathers as tools to extract their supper, they can also fashion hooks for this purpose. Indeed, in a study published in *Proceedings of the Royal Society B* ("Diversification and cumulative evolution in New Caledonian crow tool manufacture" by Gavin R. Hunt and

Russell D. Gray, April 2003, Volume 270, Issue 1717), researchers from the University of Auckland in New Zealand record that the New Caledonia crows are capable of modifying and improving their tools and also will pass these innovations on to other individuals and succeeding generations—abilities that these crows share with humans. They have also been found to be capable of "meta-tool use," using one tool on another to fulfill a task. (Additional studies on the subject of tool making by the genius crows of New Caledonia are found in *Proceedings of the Royal Society B,* May 2006, Volume 273, Issue 1590; January 2009, Volume 276, Issue 1655; May 2010, Volume 277, Issue 1686.) Perhaps it's time to give that old pejorative "bird brain" a new look—and a new meaning!

pg. 26   "work together for the common good":
To wit, a murder of crows living in my mother's suburban Philadelphia backyard (she loved them for their garrulousness, gave them gifts of bacon fat, and they would come to her call). Tired of her sauntering old tomcat coming perilously close to their nests in a grove of pine trees, the crows decided, on one of his rambles, to anoint him, en masse, with a shower of feces. The cat, crying pitifully, showed up at my mother's backdoor, covered, for want of a better word, in crap. After yet another indignation—a bath—the cat never ventured into the backyard again, preferring to sit outside on the stone patio, mocked by the cawing of the crows who, my mother was convinced, were laughing at him. (So, perhaps crows have a sense of humor, too—which is, for me, a sure sign of their intelligence.) The crows' wariness, I think, certainly in the case of humans (they will fly away even if they see me at a distance through a window) is a carefully considered skepticism—they just don't trust us. And this mistrust of humans, even their known benefactors, may be the truest indication of just how smart they are.

pg. 30   "robins nest every spring":

Blue jays, though primarily vegetarian—their love of acorns is credited for helping to spread oak trees throughout the North American continent—will eat caterpillars, beetles and other small insects. They have also been known to take the eggs and young of other birds, notably robins, earning them a rather bad reputation among us humans, who never seem to appreciate aggressive, opportunistic species such as ourselves.

CHAPTER TWO

pg. 35   "a favorite subject of song":

Miss Vogel also led us that day in a lyric that began "Listen to the bluebird gaily sing / happiness and joy to all he bring" with a chorus, "toura-loura-loura / loura-loura-loura-lou," that I would later realize actually bears some resemblance to the bluebird's own melody.

pg. 35   "a Disney cartoon caricature":

Bluebirds are featured as Cinderella's bridal couturiers and Snow White's woodland escorts—and the lyric "Mr. Bluebird's on my shoulder," indicating delight and starring an animated bluebird sporting top hat and cane, of "Zip-A-Dee-Doo-Dah" fame, is from the Academy Award-winning song from the 1946 film *The Song of the South*.

pg. 35   "smaller in stature than the robin":

The bluebird is about two-thirds of the size of a robin—six to eight inches versus its larger cousin's eight to eleven inches.

pg. 36   "nest hollows":

Bluebirds must compete for such spaces with other so-called cavity nesters, house sparrows and European starlings, non-native species that are far more aggressive than the bluebird.

pg. 36   "Thomas E. Musselman":
Sialis, "Eastern Bluebird History," www.sialis.org.

pg. 36   "marauding starlings":
Bluebird boxes have an entrance hole one-and-a-half inches
in diameter, too small for those larger birds to breech.

pg. 36   "Ralph Bell":
Sialis, "Eastern Bluebird History," www.sialis.org.

pg. 37   "American Chemical Society":
William J. Darby, "A Scientist Looks at *Silent Spring*," *Chemical and Engineering News,* 40 (1 October 1962), 60-63.

pg. 38   "major success of the environmental movement":
That said, its illegal use in the US is still suspected and its export is still allowed—endangering not just bird populations but humans, too, their food and water worldwide.

pg. 45   "snipe hunt":
Indeed, the futile snipe hunt has become synonymous with a "fool's errand" or "wild goose chase" because its quarry, the snipe, is never caught—or even glimpsed, for that matter—unless, of course, you are a "sniper," a sharpshooter so skilled that you can sight and kill a snipe, a bird that is more myth than reality.

pg. 48   "dead leaf pattern":
As described by Roger Tory Peterson, from his iconic *A Field Guide to the Birds,* Boston: Houghton Mifflin, 1980, p. 124.

pg. 50   "The Bangles":
The video's actual title, *American Woodcock: Walkius Likus Egyptianuous,* grants the woodcock a new scientific name to go along with its many nicknames.

pg. 52   "*A Sand County Almanac*":
New York: Ballantine Books, 1970, p. 33.

pg. 55   "its favorite hiding place":
It also carries a distinctive "X" marking on its back, hence its
scientific name, *Pseudacris crucifer.*

pg. 57   "helping them keep warm":
Birds have an average temperature of about 105 degrees F,
though this varies with different species. In winter, in extended cold snaps, especially when natural food sources are
covered by snow, it can be challenging for birds to maintain
their requisite body heat, hence the importance of providing
overwintering birds with high-protein, high-energy food
sources, such as fatty suet and black-oil sunflower seeds. And
what, really, is more rewarding, on a cold winter's day, than
watching birds flock to the feeders, the chickadees hopping
about, "*dee-dee-deeing*" their delight, the nuthatches and
other "upside-down" birds, such as downy and hairy woodpeckers, doing acrobatics on the suet feeders, as the shy cardinal finally gets up his courage, leading the way for his dainty
wife, finally landing on the lip of a feeder, delicately purloining
a sunflower seed, then flying off into the woods, his vibrant
red a thrill of color on a gray January day, where he will sam-
ple his prize in safety?

CHAPTER THREE

pg. 61   "usually translated":
Tim Duerden, *A History of Delaware County, New York,*
Fleischmanns, NY: Purple Mountain Press, 2007, p. 16.

pg. 61   "the people of stony country":
Richard C. Adams, *The Delaware Indians: A Brief History,*
Saugerties, NY: Hope Farm Press, 1995, p. 4. The key words

here are "usually translated." According to Herbert C. Kraft
and John T. Craft (*The Indians of Lenapehoking,* South Orange,
NJ: Seton Hall University Museum, 1985, p. 2), Lenape means
"'common' or 'ordinary' people. The term 'Lenni Lenape' is
redundant, as if to say 'the common, ordinary people.' Lenape
does not mean 'original people,' as is so often stated."

pg. 61   "present-day Margaretville":
Tim Duerden, *A History of Delaware County, New York,*
Fleischmanns, NY: Purple Mountain Press, 2007, p. 16.

pg. 61   "Lenapes' river":
The Lenapes' name for their own river was Lenape-wihit-
tuck, "stream of the Lenape" (Richard C. Adams, *The
Delaware Indians: A Brief History,* Saugerties, NY: Hope
Farm Press, 1995, p. 5).

pg. 61   "Land of the Lenape":
Herbert C. Kraft and John T. Craft, *Indians of Lenapehoking,*
South Orange, NJ: Seton Hall University Museum, 1988, p. 2.

pg. 61   "wide and fertile valley":
Tim Duerden, *A History of Delaware County, New York,*
Fleischmanns, NY: Purple Mountain Press, 2007, p. 25.

pg. 62   "from their ancestral lands":
Richard C. Adams (*The Delaware Indians: A Brief History,*
Saugerties, NY: Hope Farm Press, 1995, p. 9) attests to the
Lenapes' character: "Even after the Delawares had become
embittered and corrupted by the gross knavery of the whites
. . . and the debasing influence of alcohol, such an authority
as Gen. William H. Harrison could write these words: 'A
long and intimate knowledge of them [Delawares] in peace
and war, as enemies and friends, has left upon my mind the
most favorable impression of their character for bravery, gen-
erosity, and fidelity to their engagements.'"

pg. 63   "Fleischmanns":
          Population 342.

pg. 63   "a mile across at its widest point":
          New York State Department of Environmental Conserva-
          tion, "Pepacton Reservoir," www.dec.ny.gov.

pg. 64   "marriage of the waters":
          Jerry Cheslow, "If You're Thinking of Living in Peapack and
          Gladstone . . . ," (Real Estate), *New York Times,* August 7, 1994,
          et alia.

pg. 66   "characterized the Catskills in 1744":
          Tim Duerden, *A History of Delaware County: New York,*
          Fleischmanns, NY: 2007, p. 16.

pg. 66   "gave that river its name":
          The building of the Glen Canyon Dam, which opened in
          1963, has drastically altered the ecology of the Colorado
          River through the filtering out of sediments—used by many
          fish, amphibian, and insect species for habitat, shelter, and
          breeding—along with the resulting drop in water tempera-
          ture (and earning the ire of Edward Abbey, whose comic
          but deadly serious novel *The Monkey Wrench Gang* chroni-
          cles the misadventures of environmental activists intent on
          preventing the despoiling of the West, symbolized by the
          building of the Glen Canyon Dam). Several native fish
          species have already become extinct, and several others are
          endangered. Another effect of the dam has been to clear the
          water, and as alluring as that aquamarine green may be, it
          is not the Colorado, which means "colored red" in Spanish.
          In the summer monsoon season, when I was there, the Col-
          orado's tributaries, notably, the Little Colorado, run that
          muddy reddish-color, returning the river to at least the ap-
          pearance, if not the raging wildness, that the Mighty Col-
          orado once had.

pg. 72   "bob-bob-bobbing":
Cara Buckley, "A Little Town Near the Delaware Starts to
Shake Off the Mud and Dirt," (N.Y./Region), *The New York
Times,* July 4, 2006.

pg. 73   "damages resulted from the flooding":
USGS/FEMA, *Flood of June 26-29, 2006, Mohawk, Delaware
and Susquehanna River Basins, New York,* by Thomas P. Suro,
Gary D. Firda, and Carolyn O. Szabo, www. pubs.usgs.gov.
The *New York Times* (Fernanda Santos, "Funnel Cake,
Prized Cows, and Defiance After a Devastating Flood,"
N.Y./Region, August 18, 2006) summed up the devastation:
"The flood destroyed 35 percent of the [Delaware] county's
crops, washed away a quarter of the active cropland and
damaged 283 of its 600 farms."

pg. 73   "just a day of rain":
National Public Radio, Morning Edition, "New York Town
Isolated by Rising Water," reported by Steve Inskeep, June
29, 2006, www.npr.org.

CHAPTER FOUR

pg. 80   "approximately 750 species":
The Lepidopterists' Society, "Lepidopteran Information,"
www .lepsoc.org.

pg. 80   "150,000 species of Lepidoptera":
"Lepidoptera" derives from the Greek, "lepidos" (scale) and
"pteros" (wing), connoting the only insects with scales covering most of the wings and some of the body. The scales
serve a largely protective purpose, overlapping each other
like shingles, their color providing camouflage and temperature regulation, darker hues soaking up sunlight, helping
keep the insect warm.

pg. 81   "only 20,000 are butterflies":
The Lepidopterists' Society, Lepidopteran Information, www.lepsoc.org.

pg. 81   "some 130,000 species":
The Lepidopterists' Society, Lepidopteran Information, www.lepsoc.org.

pg. 81   "moths":
According to the Lepidopterists' Society, in the Nearctic eco-zone, which includes the US, Canada, and parts of Northern Mexico, there are some 10,850 moth species, vastly outnumbering their highfalutin cousins.

pg. 81   "alder, elm, cherry, and birch":
New York State Department of Environmental Conservation website, "Tent Caterpillars," www.dec.ny.gov. These outbreaks occur in cycles—in New York, they tend to occur in ten-year intervals, lasting, on average, three years. It is indeed shocking, to look up in spring and see wide black swaths on the mountainsides where entire stands of trees have been defoliated by these caterpillars. Most healthy trees can survive such an onslaught—they will re-leaf in July—even for several seasons. But Nature also sends reinforcements. Such eruptions spark a surge in the "friendly fly" population—small black flies so called because they are attracted to people's skin (probably for the salt), also giving them the unappetizing name of "flesh flies." They do not bite, however, though they are a nuisance. But they do serve a purpose—and it is in its adaptations that Nature is at its most sublime. Friendly flies attack the cocoons of forest tent caterpillars, eating the larvae. The subsequent years of an outbreak are less severe, due to these flies, whose numbers collapse when the caterpillar infestation is over. Another name for the friendly fly is the "government fly," as many Catskillers, including friends who have made this claim to

me, believe that the "government" releases these flies for this purpose. I have even been told that each fly costs the "government" fifty cents apiece! For the record, this is a naturally occurring phenomenon—the flies are native to New York State. Nature, as always, is far more competent, and certainly more economic and elegant in its solutions, than any human construct.

pg. 83   "when it is finally warm":
Southern climes may feature as many as three generations through summer.

pg. 85   "rosy maple during the day":
Ohio Birds and Biodiversity (*Colorful Camouflage, Rosy Maple Moths,* by Jim McCormac),
www.jimmccormac.blogspot.com.

pg. 87   "eyespots":
It had long been theorized, even accepted, that such "eyespots" were also meant to mimic the eyes of the predator's own enemies, such as raptors, which would work to intimidate smaller birds. But when Martin Stevens, a behavioral ecologist, and his associates at Cambridge University put this notion to the test, they concluded that it is the "visual loudness" of the markings, not necessarily their resemblance to an "eye," that deters predators (Martin Stevens et al., "Conspicuousness, not eye mimicry, makes 'eyespots' effective antipredatory signals," *Behavioral Ecology,* 2008, Volume 19, Issue 3, pp. 525-531).

pg. 87   "quarter moons":
These moonlike markings may give the Luna, which is the Latin word for "moon," its name, though the Luna's use of the moon as a navigational guide is also a credible explanation.

pg. 87   "the fact that several of its members":
Other imposing "eyespot" Saturniids that are also Catskill
denizens, and ones which I have been privileged to see, include the Polyphemus and Io. They deserve their mythical
monikers. The Polyphemus, with a wingspan of around
four to six inches, is named for the giant Cyclops of the same
name, who had a large round eye in the center of his fore-
head, which the hero Odysseus disabled, targeting the "eye"
much as predators seek out the moth's eyespots on its reddish to yellow-brown hindwings (fortunately, with less crip-
pling effect). Io was a beautiful young woman beloved of
Zeus. The moth that bears her name, though smaller than
other Saturniids (with wings averaging two to three inches
wide), is just as striking, boasting prominent dark eyespots,
also on the hindwings. Io may also derive from "Eos," the
Greek goddess of dawn, as the moth's colors—yellow, orange, and red—are reminiscent of the soft glow of a newly
breaking day. These moths and some other Saturniids
(such as the Promethea, named for the Titan Prometheus,
who stole fire from Mount Olympus and gave it to humankind, a deed perhaps remembered in its smoky reddish-
brown wings that seem to smolder in dim light), are as
impressive, in name as well as appearance, as the most beautiful butterflies.

pg. 87   "the planet Saturn":
J. A. Powell, "Lepidoptera (Moths, Butterflies)," in V. H.
Resh and R. T. Cardé (editors), *Encyclopedia of Insects,* San
Diego, CA: Elsevier Science (USA), 2003, pp. 661-663.
William Leach posits in *Butterfly People: An American En-
counter with the Beauty of the World* (New York: Vintage
Books, 2014, p. 98) that Saturniidae derives from Saturnia,
the Roman goddess also known as Juno, daughter of Saturn.

pg. 88   "distinguish many moths from butterflies":
Butterflies feature long, slender antennae topped by tiny

knobs, sensory receptors that enable them to find food and a mate, migrate, and even tell time.

pg. 90 "bat-moth interactions":
Jesse R. Barber et al., "Moth tails divert bat attacks: Evolution of acoustic deflection," *PNAS* (*Proceedings of the National Academy of Sciences*), March 3, 2015, Volume 112, Number 9, pp. 2813-2816.

pg. 92 "no sign of slowing down":
At the end of the 2014-15 hibernating season, bats with WNS were confirmed in twenty-six states and five Canadian provinces. Center for Biological Diversity / White-nose Syndrome.org.

pg. 92 "$53 billion a year":
J. G. Bowles et al., "Economic importance of bats in agriculture," *Science,* April 1, 2011, Volume 332, Number 6025, pp. 41-42.

CHAPTER FIVE

pg. 96 "Catskill farm country":
Naturalist and hawk expert Peg DiBenedetto, from an old Catskill farming family, offered me this insight: "Sharp-topped walls were built for the sheep; there was minimal if any goat presence here, and that design would not have been effective against goat breakouts anyway."

pg. 98 "sandstone in the Western Catskills":
"Bluestone" is a term used to describe the feldspathic sandstone that is still mined today in the Western Catskills, whether that stone is the dark blue-colored stone of the High Peaks or the grayish "bluestone" of Delaware County.

pg. 98   "heights once touched 28,000 feet":
             Alf Evers, Robert Titus, and Tim Weidner, *Catskill Mountain Bluestone* (Robert Titus, "The Bluestones of Ulster County"), Fleischmanns, NY: Purple Mountain Press, 2008, p. 41.

pg. 100   *"wojak"*:
             Pritchard, Evan T., *Native New Yorkers: The Legacy of the Algonquin People of New York,* San Francisco, CA: Council Oak Books, 2002, p. 376.

pg. 108   "The proportion of 'waste' to 'useful product' is ten to one":
             "Bluestone"—New York State Department of Environmental Conservation, www.dec.ny.gov.

pg. 110   "Department of Environmental Conservation":
             "Red Fox"—New York State Department of Environmental Conservation, www.dec.ny.gov.

pg. 111   "for thirty million years":
             J. David Henry, *Red Fox: The Catlike Canine,* Washington, D.C.: Smithsonian Books, 1996, p. 27.

pg. 116   "the fox's stalking":
             J. David Henry, *Red Fox: The Catlike Canine,* p. 71.

pg. 116   "a trait it also shares with cats":
             J. David Henry, *Red Fox: The Catlike Canine,* p. 71.

pg. 116   "bona fide member of the Family *Canidae*":
             J. David Henry, *Red Fox: The Catlike Canine,* p. 70.

pg. 116   "both physical and behavioral":
             J. David Henry, *Red Fox: The Catlike Canine,* p. 70.

pg. 116   "lateral threat display":
             J. David Henry, *Red Fox: The Catlike Canine,* pp. 70-72.

pg. 119 "It was standing in the middle of the road":
My use of "it" to describe the fox is only because I didn't know its sex. Dog foxes are bigger than vixens, but the difference is imperceptible, certainly when seeing one alone. I suspected the fox was male because it was out and about in summer, probably more than a vixen with kits would have been. But I couldn't be sure—and this lack of gender differentiation perhaps goes again to the fox's in-between-ness.

CHAPTER SIX

pg. 133 "sighted in the Catskills region":
A Birdseye View of the Great Northern Catskills (Birdwatching: Greene County, NY) by Richard Guthrie, 2006, www.greatnortherncatskills.com.

pg. 133 "largest in North America":
The pileated, sixteen to nineteen inches in length, is second in size only to the ivory-billed woodpecker, which measures twenty inches in length. This native of southeastern woodlands is generally regarded as extinct, though the Cornell Lab of Ornithology reports credible sightings as recently as 2005, and the U.S. Fish & Wildlife Service has released "The Ivory-Billed Woodpecker Recovery Plan" "wherever and whenever it is needed." (www .birds.cornell.edu)

pg. 133 "dipped in raspberry juice":
Roger Tory Peterson, *A Field Guide to the Birds,* Boston: Houghton Mifflin, 1980, p. 270.

pg. 135 "What ceases for most birds":
A notable exception is the northern cardinal. Both male and female sing year-round, and though both molt, the male retains his bright red feathers and the female also keeps her pretty, though more modest garb. Few sights lift one's spirits as a car-

dinal pair at the feeders on a cold, snowy day in the Catskills, their sweet whistling defying the silence of a winter day.

CHAPTER SEVEN

pg. 143   "Catskil Mountains":
Alf Evers, *The Catskills: From Wilderness to Woodstock,* Woodstock, NY: The Overlook Press, 1982, p. 531.

pg. 144   "I had never been on the Catskills":
Andrew Burstein, *The Original Knickerbocker: The Life of Washington Irving,* NY: Basic Books, 2007, p. 117.

pg. 144   "Thomas Cole":
Durand, Gifford, and Cole each painted one of the most iconic and spectacular features of the wild Catskills, "Kaaterskill Clove," a gorge as deep as 2,500 feet in the Eastern Catskills, formed by glaciations and the erosive effects of the Kaaterskill Creek, a tributary of Catskill Creek. The Clove houses, among other waterfalls, Kaaterskill Falls, which plunges 231 feet and still takes one's breath away.

pg. 145   "behind his native dikes":
T. Morris Longstreth, *The Catskills* ("Origin of the Name Catskill, one man's opinion"), 1918 (reprinted by Hope Farm Press, www.hopefarm.com, copyright 2002).

pg. 145   "the name 'Catskills' derives":
Horatio Stafford (*Gazeteer,* 1813, pp. 78-79) stated that the Dutch word for a domestic she cat   was "kat," and that "kater" signified a tomcat. Hence, "Kaaterskill" meant "tomcat's-creek," "Catskill" meant "she-cat's-creek," and "Catskill Mountains" meant "she-cat's-creek mountains." Evers (op. cit., p. 533), who references these findings, reports that these etymologies were, at the time, generally accepted.

pg. 145 "John James Audubon":
The first plate in John James Audubon's The *Viviparous Quadrupeds of North America* (1842) is of the "American wildcat," as Audubon termed it, which we know today as the bobcat, painted in all its beauty, ferocity and defiance.

pg. 147 "no reason to fear—humans":
J. David Henry, *Red Fox: The Catlike Canine*, Washington, DC: Smithsonian Books, 1996, p. 23: "The red foxes found within Prince Albert National Park have not been trapped or hunted for more than fifty years and as a result have lost their shyness of humans."

pg. 148 "enjoying the day":
The bobcat did not appear rabid, which is often characterized by symptoms that include erratic movements, aggressive behavior, and drooling or "foaming at the mouth." Still, it is a question that needs to be considered when certain species vulnerable to rabies, such as bobcat and fox, skunk and raccoon, approach humans or their dwellings in daylight hours.

pg. 148 "social behavior":
Kevin Hansen, *Bobcat: Master of Survival,* New York: Oxford University Press, 2007, p. 32.

pg. 149 "Department of Environmental Conservation":
Bobcats: New York State Department of Environmental Conservation, www.dec.ny.gov. According to NYS DEC, male bobcats are around one-third larger than their female counterparts. Both sexes can weigh more than thirty pounds; averages for males and females are twenty-one and fourteen pounds respectively. The body length for males averages thirty-four inches; thirty inches is the average length for females. The tail length is five to six inches for both sexes. Bobcats can stand at the shoulder sixteen to twenty-four inches.

pg. 149   "He can lick his weight in wildcats":
Hansen, *Bobcat,* p. 10.

pg. 151   "bobcats and lynx are even less in evidence":
Jerry Kobalenko, *Forest Cats of North America,* Richmond
Hill, Ontario: Firefly Books, 1997, p. 80. Hansen (*Bobcat*)
notes the bobcat figures prominently in Navajo creation
mythology (p. 4); the Navajos also believed that bobcats had
the ability to "witch" people with their whiskers, which had
special power (p. 103).

pg. 151   "small mammals, birds, and deer"
Kobalenko, *Forest Cats*, p. 22.

pg. 152   "several northern U.S. border states":
U.S. Fish and Wildlife Service (USFWS) can confirm the
presence of lynx in Maine, Montana, Washington, and Colorado. Never populous in the Lower 48, the lynx was nearly
extirpated due to trapping and habitat loss. In 2000, the
Canada lynx was declared a threatened species and is now
federally protected from human exploitation by the Endangered Species Act.

pg. 152   "*Lynx rufus*":
"Lynx," a mysterious-sounding name that befits a mysterious
cat, is reasonably said to derive from the Latin "lux," (light)
and/or the Greek "leukos," meaning white, though Hansen
(*Bobcat,* p. 11) states: "Lynx in Latin means lamp, from an original Greek word meaning 'to shine.' Both refer to the reflective
quality [see *tapetum lucidum*, Hansen, p. 30] of the cat's eyes
when struck by light at night."

pg. 152   "imitating its prey":
Kobalenko, *Forest Cats,* p. 32.

pg. 153   "the lynx's disappearance":
Harold Faber, "After 100-Year Absence, Shy Lynx Is Returning," *The New York Times,* April 2, 1987.

pg. 154   "breeding population of lynx in New York":
Lynx: New York State Department of Environmental Conservation, www.dec.ny.gov. The Wildlife Conservation Society
(of the Bronx Zoo) also conducted surveys in the High Peaks,
from 1998 to 1999, in search of lynx, but no evidence of Canada
lynx was found.

pg. 155   "little black panthers":
"Black panthers" don't exist as a separate species. The term
describes any large wild cat with a black coat. In South
America, where this cat lives in the Western Hemisphere, it
is a jaguar with a melanistic gene; in Asia and Africa, it is a
leopard. But in the Catskills, despite the insistence on "black
panthers," what people are probably seeing is a large black
house cat, suitable for Halloween, or perhaps a fisher, another
critter that has made its way back to the Catskills, often called
a "fisher cat."

pg. 155   "mountain lions in this area since 2000":
Christopher Ketcham, "Cat Eyes," *Earth Island Journal,* summer 2011, www.earthisland.org.

pg. 155   "then it's a mountain lion":
Ketcham, "Cat Eyes," *Earth Island Journal,* summer 2011.

pg. 156   "twenty-six to thirty-two inches long":
Eastern cougar: New York State Department of Environmental Conservation, www.dec.ny.gov.

pg. 156   "its trademark tail, is five to nine feet":
Eastern cougar: NYS DEC, www.dec.ny.gov.

pg. 157   "same habitat as its chief prey":
Kobalenko, *Forest Cats,* p. 10. "Even subtle color differences
between cougar populations seem to correspond to the shad-
ings of indigenous deer," Kobalenko adds.

pg. 157   "Columbia County":
Cougar (Mountain Lion): *Catskill Mountaineer,*
www.catskillmountaineer.com.

pg. 157   "Eastern Cougar Sightings":
Eastern cougar: NYS DEC, www.dec.ny.gov.

pg. 158   "open debate, review and public commentary":
John Leahy, "Mountain Lions in Eastern New York State,"
Leahy Institute (originally published in the Eastwick Press,
2006), www.leahyinstitute.org.

pg. 158   "the feasibility of this is much debated":
In his 2013 study, wildlife biologist John Laundre argued that
the Adirondack Park could support up to 350 cougars, a theory debunked by Dr. Rainer Brock who, in his 1981 study,
posited that cougars cannot live in close proximity to people
and roads, an idea that, in turn, Laundre has dismissed. John
Laundre, "The feasibility of the north-eastern USA support-
ing the return of the cougar *Puma concolor*," *Oryx: The Inter-
national Journal of* Conservation, Vol. 47, Issue 1, January
2013, pp. 96-104.

pg. 158   "Milford, Connecticut in June 2011":
The story of the long-distance mountain lion, as well as the
cougar's history and prospects, is beautifully and com-
pellingly told in *Heart of a Lion: A Lone Cat's Walk Across
America* by William Stolzenburg (New York: Bloomsbury
USA, 2016).

pg. 159   "a world that is out of kilter":
Michael Robinson, Center for Biological Diversity, "With Tolerance, Apex Predators Like Cougars Can Return, Restore Ecosystems," June 16, 2015, www.biologicaldiversity.org.

CHAPTER EIGHT

pg. 165   "true original Native of America":
According to the United States Diplomacy Center, as referenced in an article appearing on Smithsonian.com (Jimmy Stamp, *American Myths: Benjamin Franklin's Turkey and the Presidential Seal,* January 25, 2013, www.smithsonianmag.com), the "myth" of Franklin advocating for the turkey as the National Bird is "entirely false." That said, in an oft-quoted letter to his daughter (The Franklin Institute, Benjamin Franklin/FAQ, www.fi.edu), Ben does seem to prefer it, praising the turkey which "would not hesitate to attack a Grenadier of the British Guards who should presume to invade his Farm Yard with a red coat on," while he maligns the bald eagle as a "bird of bad character," "lazy" and a "coward," "by no means a proper [national] emblem."

pg. 166   "full of a wild, harmless look":
John Burroughs, *In the Catskills,* Fleischmanns, NY: Purple Mountain Press, 2010, p. 15.

CHAPTER NINE

pg. 177   "*Peterson Field Guide to Mammals of North America*":
Fiona Reid, Boston: Houghton Mifflin Harcourt, 2006, p. 450.

pg. 178   "Wildlife biologist Larry Bifaro":

Bifaro, who lived on the other side of the mountain, had seen the bear when I first did, hypothesizing that she was female because of her smaller size and expressing concern that she was "a bit too bold," traversing backyards with ease, where she had probably first encountered bird feeders. "Let's hope she doesn't stick her nose in the wrong feeder," he'd said, adding that "nuisance" bears, ones that get too used to humans and their food, are often shot, even out of hunting season.

pg. 179   "sprinting for the woods":

"Sloth," a term still used to describe a pack or group of bears, which derives from the Middle English *slowth,* meaning "slow," is clearly a misnomer—the large "lumbering" black bear can run at speeds exceeding 25 mph.

pg. 179   "males top the scales at 200 to 600 pounds":

There are credible variations recorded in the height, length and weight of black bears, as well as other of its characteristics. The statistics given here are drawn from information provided by the website of the New York State Department of Environmental Conservation—www.ny.dec.gov (which profiles black bear as well as other mammals native to New York)— because my subject is black bears in the Catskills. The website of the North American Bear Center, in Ely, Minnesota (www.bear.org), which offers bear facts while dispelling bear myths and provides comprehensive, up-to-date information, describes the black bear as having a length of as much as *seven feet* "from nose to tail," and reports that "lean bears can exceed *30 mph*." They grow 'em big—and fast—in Minnesota!

pg. 180   "don't turn your back on it":

That dictum of bear country, "Never run away from a black bear, lest you elicit the carnivore's chase behavior," is refuted by wildlife biologist Lynn Rogers, who is quoted in *The Great*

American Bear (Jeff Fair and Lynn Rogers, Minocqua, WI: NorthWord Press, 1990, p. 51) as saying "I've always heard this, but never from anyone who's tested it. Haven't seen it myself." However, as recently as September 2014, a Rutgers University student, hiking in New Jersey, was attacked and killed when he and four friends "scrambled to get away from the bear, all running in different directions" (*New York Times,* "Black Bear Kills Rutgers Student During a Hike in New Jersey" by Tatiana Schlossberg (N.Y./Region), September 24, 2014, which also notes this was "the state's first recorded death linked to a black bear.") Documented black bear attacks are indeed rare. But that particular "bear country dictum" is not one that I would test.

pg. 180   "lord of the wild Catskills":
Alf Evers, *The Catskills: From Wilderness to Woodstock,* Woodstock, NY: The Overlook Press, 1982, p. 304.

pg. 182   "two stones for every dirt":
As my neighbor says, "Here in the Catskills, we garden with a pickaxe."

pg. 183   "nearly 300,000 (287,000) acres":
By contrast, the Adirondack Forest Preserve weighs in at a hefty 2.6 million acres.

pg. 184   "'place of fish' in the Lenape language":
NYS Department of Environmental Conservation, "Ashokan Reservoir," www.dec.ny.gov.

pg. 187   "the bear, wiser than I, wanted to avoid":
Indeed, especially among grizzly bears, a significantly more aggressive species, cameras with large photographic lenses, perceived by the bears as huge, staring eyes, have been known to trigger attacks on photographers who are not aware that their "shooting" has been interpreted as threatening behavior.

pg. 187   "they have yet to hear their first growl":
The North American Bear Center, "Do black bears growl?"
www.bear.org.

pg. 189   "a large brain compared to their body size":
Some wildlife biologists—notably Benjamin Kilham, author of *Out on a Limb: What Black Bears Have Taught Me
About Intelligence and Intuition,* whose work in the New
Hampshire woods with black bears over more than twenty
years has led to the startling and controversial (to some)
thesis that bears, usually regarded as solitary, are indeed
social and cooperative, even sometimes displaying altruistic
behavior, while likening bear society, in the animals' mu-
tually beneficial interactions, to those of primitive humans—believe that their intelligence is comparable to that
of the great apes. Clearly, we have so much still to learn
about this critter.

pg. 189   "made her look less fierce, endearing even":
The term "black bear" is actually a misnomer in that not
all black bears are black. Our Catskill black bear, among
the first glimpsed by Europeans when they arrived in
America, specifically the East, may have indeed prompted
the name. But as settlers moved west, they encountered
"black bears" that were chocolate, cinnamon, red, blond,
blue gray and even white—same bear, but with different
coat colors, called "phases" or "morphs." Jeff Fair (*The
Great American Bear,* Minocqua, WI: NorthWord Press,
1990, p. 63) notes that brown color phases, occurring more
commonly in the West, "may serve as protective coloration
(mimicking the grizzly bear) or to absorb less solar heat
under more-open forest cover." Black fur absorbs heat most
readily, helping bears living in cool climes such as the
Catskills stay warm. The blue-gray coat of the striking
"glacier" bear, rarer in recent years due to inbreeding with
"black" black bears, allows it to virtually melt into the

frozen landscape of its subarctic home. The white or cream-colored Kermode bear, native only to a few islands off the coast of British Columbia, often mistaken for an albino, though in the minority of that specific population, is in fact another color morph of the black bear. Named "spirit" bears by Native Americans, these hauntingly beautiful, increasingly rare bears figure prominently, as bears do in many indigenous cultures, in local mythology and lore. There can also be a variation, in a minority of the "black" black bear population, of a white blaze or "V" on the chest. My own Catskill black bears, however, are not so sporty, preferring their daytime and evening dress to be classic basic black.

pg. 189   "accustomed to being in people's backyards":
          "To feed a bear is to kill it" is another well-known "bear country dictum" (articulated by Dave Taylor in *Black Bears: A Natural History*, Markham, Ontario: Fitzhenry and White-side, 2006, p. 99), which is ascribed to by many, including NYS DEC. The argument is that feeding bears, even inadvertently (e.g., bird feeders), can turn them into "nuisance bears," which can get bears shot, for starters, or make them lose their fear of people, which can cause them to be aggressive, with potentially dangerous consequences to both bears and humans. Wildlife biologist Lynn Rogers, whose work with black bears Jane Goodall has called "one of those rare long-term studies where each successive year make the whole that much more valuable," has sparked controversy by hand-feeding bears in order to observe them in their natural setting. Whether feeding them compromises how the bears act, and whether the bears have become "nuisances," as a result, to surrounding communities are subjects of debate. But regardless of one's viewpoint (and I do *not* espouse feeding black bears in the Catskills, for any reason), Rogers' "walking with bears," his close encounters with them as a way of learning about them in the wild, has helped dispel ignorance about

these critters, and his researches have significantly augmented our understanding of the black bear.

pg. 191   "helping hunters find their way home"
Alf Evers, *The Catskills: From Wilderness to Woodstock,*
Woodstock, NY: The Overlook Press, 1982, p. 301.

pg. 194   "attributed to the Evil Spirit":
Joshua B. Lippincott (compiler), *Folk-lore and Legends of
the North American Indians* (London: Abela Publishing,
2009), p. 115. This singular—and invaluable—assemblage
of Native American legends and lore was first published by
J. B. Lippincott in 1891.

pg. 194   "participants in Europe, Asia and America":
Alf Evers, *The Catskills: From Wilderness to Woodstock,*
Woodstock, NY: 1982, p. 302.

pg. 194   "Switzerland":
Alf Evers, *The Catskills: From Wilderness to Woodstock,*
Woodstock, NY: 1982, p. 302.

pg. 194   "conflict adverse":
All Animals, "Bear Country" by Karen E. Lange. September
/October 2011, P. 28.

pg. 195   "especially when it comes to humans":
All Animals, "Bear Country," by Karen E. Lange. September
/October 2011, p. 28.

pg. 195   "spiders, dogs and lightning":
All Animals, "Bear Country," by Karen E. Lange. September
/October, p. 27.

pg. 195   "vending machine accidents":
All Animals, "Bear Country," by Karen E. Lange. September
/October, p. 27.

pg. 195   "child's cuddly companion":

In 1902, President Theodore Roosevelt, celebrated big-game hunter and conservationist, refused to shoot a wounded black bear tied to a tree during a hunting trip to Mississippi, a scene *Washington Post* cartoonist Clifford K. Berryman lampooned with the caption "Drawing the line in Mississippi," featuring Roosevelt standing, in his trademark Rough Riders outfit, with his back to the guide and captive bear, his rifle at rest in his right hand as he rejects the unsportsmanlike scene with his left. (According to the website of the Theodore Roosevelt Association, www.theodoreroosevelt.org, Roosevelt couldn't stomach shooting the injured bear, which had been attacked by the guide's dogs, and instructed him to put the bear out of its misery—not exactly how this beloved story has come down through the years.) Berryman continued to portray the bear as TR's mascot in his subsequent cartoons, with the adult bear now transformed into a jaunty little cub accompanying Roosevelt at home and abroad, thus solidifying the connection between Teddy and his bear. Shortly after the original cartoon was published (November 16, 1902), Morris Michtom placed two stuffed toy bears his wife had made in the window of his Brooklyn, New York candy store, calling them "Teddy's bears," which politician Roosevelt was only too happy to endorse. Michtom profited too—the bears became so popular that he soon began to mass-produce them and would go on to found the Ideal Novelty and Toy Company, thanks largely to the success of "Teddy's bear" or as we know it today, the teddy bear.

pg. 196   "wild, yet humanlike":

A friend who is also a deer hunter in these Catskill foothills only hunted bear once. He told me he was "shocked" when he saw the skinned body of the bear he had shot. "It looked so human," he said, "I knew right then, I could never hunt bear again."

pg. 196   "The Catskill Mountains, 1867":
Referenced in Alf Evers, *The Catskills: From Wilderness to Woodstock,* Woodstock, NY: The Overlook Press, 1982, p. 303.

pg. 197   "many centuries earlier":
Alf Evers, *The Catskills: From Wilderness to Woodstock,* Woodstock, NY: The Overlook Press, 1982, p. 302.

pg. 200   "way overblown":
Jeff Fair and Lynn Rogers, *The Great American Bear,* Minocqua, WI: NorthWord Press, 1990, p. 48. Stephen Herrero, author of *Bear Attacks: Their Causes and Avoidance,* referenced in the same source, agrees: "[Black bear] mothers may bluff ferociously when cubs are endangered, but rarely attack."

pg. 200   "when protecting their young":
According to the website of the North American Bear Center, in Ely, Minnesota (www.bear.org), 70 percent of the killings by grizzly bears are by mothers defending cubs.

pg. 201   "they just seem to enjoy them":
North American Bear Center, "Why do bears play?" www.bear.org.

Resources

BOOKS

Field Guides

The Birder's Handbook: A Field Guide to the Natural History of North American Birds, Paul Ehrlich, David S. Dobkin, Darryl Wheye, Simon and Schuster: New York, 1988.

A Field Guide to the Grand Canyon, Stephen R. Whitney, Seattle, WA: The Mountaineers, 1996.

National Audubon Society Field Guide to North American Birds: Eastern Region, National Audubon Society, New York: Alfred A. Knopf, 1994.

National Audubon Society Field Guide to North American Butterflies, Robert Michael Pyle, New York: Alfred A. Knopf, 1981.

National Audubon Society Field Guide to North American Insects and Spiders, National Audubon Society, New York: Alfred A. Knopf, 1980.

National Audubon Society Field Guide to North American Mammals, John O. Whitaker & National Audubon Society, New York: Alfred A. Knopf, 1996.

National Audubon Society Field Guide to North American Reptiles and Amphibians, John L. Behler and F. Wayne King, New York: Alfred A. Knopf, 1979.

National Audubon Society Field Guide to North American Trees: Eastern Region, National Audubon Society, New York: Alfred A. Knopf, 2002.

National Audubon Society Field Guide to North American Wildflowers: Eastern Region, National Audubon Society, New York: Alfred A. Knopf, 1980.

New York Wildlife Viewing Guide, Watchable Wildlife, Cambridge, MN: Adventure Publications, 2012.

Peterson Field Guide to Eastern Birds, Roger Tory Peterson, Boston: Houghton Mifflin, 1980.

Peterson Field Guide to Eastern Butterflies, Paul A. Opler, Boston: Houghton Mifflin Harcourt, 1998.

Peterson Field Guide to Eastern Trees, George A. Petrides, Boston: Houghton Mifflin Harcourt, 1998.

Peterson Field Guide to Insects: America North of Mexico, Donald J. Borror and Richard E. White, Boston: Houghton Mifflin Harcourt, 1998.

Peterson Field Guide to Mammals of North America, Fiona Reid, Boston: Houghton Mifflin Harcourt, 2006.

Peterson Field Guide to Moths of Northeastern North America, David Beadle and Seabrooke Leckie, Boston: Houghton Mifflin Harcourt, 2012.

Peterson Field Guide to Reptiles and Amphibians: Eastern and Central North America, Joseph T. Collins and Roger Conant, Boston: Houghton Mifflin Harcourt, 1998.

Peterson Field Guide to Wildflowers: Northeastern and North-central North America, Margaret McKenney, Boston: Houghton Mifflin Harcourt, 1998.

The Sibley Field Guide to Birds of North America, David Allen Sibley, New York: Alfred A. Knopf, 2003.

Smithsonian Field Guide to the Birds of North America, Ted Floyd, New York: Harper Perennial, 2008.

Texts

Abbey, Edward, *Desert Solitaire,* New York: Touchstone, 1968.

Adams, Richard C., *The Delaware Indians: A Brief History,* Saugerties, NY: Hope Farm Press, 1995.

Ashby, Eric, *My Life with Foxes,* London: Robert Hall, 2000.

Burroughs, John, *Catskills and Hudson Valley Essays,* Wickford, RI: New Street Communications: 2013.

Burroughs, John, *In the Catskills,* Fleischmanns, NY: Purple Mountain Press, 2010.

Burroughs, John, *Signs and Seasons,* Syracuse: NY: Syracuse University Press, 2006.

Carson, Rachel, *Silent Spring,* Boston: Houghton Mifflin, 2002.

Craft, Herbert C. and John T. Craft, *Indians of Lenapehoking,* South Orange, NJ: Seton Hall University Museum, 1988.

Dillard, Annie, *Pilgrim at Tinker Creek,* New York: Harper Perennial Modern Classics, 2007.

Duerden, Tim, *A History of Delaware County, New York,* Fleischmanns, NY: Purple Mountain Press, 2007.

Eiseley, Loren, *The Immense Journey: An Imaginative Naturalist Explores the Mysteries of Man and Nature,* New York: Vintage, 1959.

Emerson, Ralph Waldo, *Nature and Selected Essays,* New York: Penguin Classics, 2003.

Evers, Alf, *The Catskills: From Wilderness to Woodstock,* Woodstock, NY: The Overlook Press, 1982.

Evers, Alf, *In Catskill Country: Collected Essays on Mountain History, Life and Lore,* Woodstock, NY: The Overlook Press, 1995.

Evers, Alf, Robert Titus and Tim Weidner, *Catskill Mountain Bluestone,* Fleischmanns, NY: Purple Mountain Press; Schenectady, NY, 2008: New York Folklore Society, 2008.

Fair, Jeff and Lynn Rogers, *The Great American Bear,* Minocqua, WI: NorthWord Press, 1990.

Garber, Steven D., *The Urban Naturalist,* New York: John Wiley & Sons, 1987.

Henry, J. David, *How to Spot a Fox,* New York: Mariner Books, 1993.

Henry, J. David, *Living on the Edge: Foxes,* Minocqua, WI: Wildlife Series, 1996.

Henry, J. David, *Red Fox: The Catlike Canine,* Washington, DC: Smithsonian Books, 1996.

Kilham, Benjamin, *Out on a Limb: What Black Bears Have Taught Me About Intelligence and Intuition,* White River Junction, VT: Chelsea Green Publishing, 2013.

Lippincott, Joshua B. (compiler), *Folk-lore and Legends of the North American Indians,* London: Abela Publishing, 2009.

Leach, William, *Butterfly People: An American Encounter with the Beauty of the World,* New York: Vintage Books, 2014.

Leopold, Aldo, *A Sand County Almanac,* New York: Ballantine Books, 1970.

Marsh, George Perkins, *Man and Nature, Or, Physical Geography as Modified by Human Action,* Seattle, WA: University of Washington Press, 2003.

Merton, Thomas, *When the Trees Say Nothing: Writings on Nature,* edited by Kathleen Deignan, Notre Dame, IN: Sorin Books, 2003.

Pritchard, Evan T., *Native New Yorkers: The Legacy of the Algonquin People of New York,* San Francisco, CA: Council Oak Books, 2002.

Sive, Mary Robinson, *Lost Villages,* Delhi, NY: Delaware County Historical Association, 1998.

Stradling, David, *Making Mountains: New York City and the Catskills,* Seattle, WA: University of Washington Press, 2007.

Taylor, Dave, *Black Bears: A Natural History,* Markham, Ontario: Fitzhenry and Whiteside, 2006.

Thoreau, Henry David, *The Journal: 1837-1861,* New York: New York Review of Books, 2009.

Thoreau, Henry David, *Walden or, Life in the Woods,* Mineola, NY: Dover, 1995.

Titus, Robert, *The Catskills: A Geologic Guide,* Fleischmanns, NY: Purple Mountain Press, 2004.

Van Valkenburgh, Norman J., Old *Stone Walls: Catskill Land and Lore,* Fleischmanns, NY: Purple Mountain Press, 2004.

Whitman, Walt, *Leaves of Grass,* New York: Vintage, 1992.

WEBSITES

Audubon Guide to North American Birds, www.audubon.org/field-guide.

A Birdseye View of the Great Northern Catskills (Birdwatching: Greene County, NY), by Richard Guthrie, 2006, www.greatnortherncatskills.com.

Butterflies and Moths of North America, www.butterfliesandmoths.org.

Catskill Mountaineer, www.catskillmountaineer.com.

Catskill Mountainkeeper, www.catskillmountainkeeper.org.

The Cornell Lab of Ornithology, All About Birds, www.allaboutbirds.org.

The Cornell Lab of Ornithology, www.birds.cornell.edu.

eBird, www.ebird.org.

eNature Field Guides (Amphibians/Birds/Insects/Mammals/Reptiles
     /Trees/Wildflowers), www.enature.com.

Learn About Butterflies, www.learnaboutbutterflies.com

The Lepidopterists' Society, www.lepsoc.org.

MyWildflowers.com, www.mywildflowers.com.

National Audubon Society, www.audubon.org.

National Geographic, Identification and Information, (Amphibians/
     Birds /Bugs/Mammals/Reptiles), animals.nationalgeographic.com.

National Wildlife Federation, www.nwf.org.

New York City Department of Environmental Protection (Cannonsville
     /Pepacton Reservoirs), www.nyc.gov.

New York State Department of Environmental Conservation (Animals
     /Plants and Aquatic Life/Birds/Lakes and Rivers/Lakes and
     Waters), www.dec.ny.gov.

North American Bear Center, www.bear.org.

Sialis, "A Resource for Helping Bluebirds and Other Native Cavity-
     nesters," www.sialis.org.

WildlifeNorthAmerica.com (North American Mammals/Birds/Reptiles
     /Insects), www.wildlifenorthamerica.com.

Wildflower Search (Wildflower Identification Website),
     www.wildflowersearch.com.

Wildlife Research Center, www.bearstudy.org.

Acknowledgments

ANY PEOPLE HAVE CONTRIBUTED GENEROUSLY TO THE making of *The Quarry Fox and Other Critters of the Wild Catskills*. First and foremost, I wish to thank Jennifer Lyons, my literary agent extraordinaire and friend, who truly cares for writers and writing. At The Overlook Press, thanks to legendary Catskiller and publisher, Peter Mayer, for believing in the book; Allyson Rudolph, who was the first to love it; Tracy Carns, whose steady and creative hand guided it along the way; and thanks to Chelsea Cutchens, for her caring and careful editing and smart, sensitive suggestions. Thanks to Writers in the Mountains, especially my WIM colleagues and friends, Catskills culture maven Simona David and esteemed poet Sharon Israel Cucinotta, and WIM's radio program, "The Writer's Voice" (WIOX, Roxbury, NY), notably host Annie Hersh, which was the first to give these Catskills critters a voice. Thanks, also, to Jack McShane and Peg DiBenedetto, my friends and fellow naturalists, both of whom know more about the Catskills than I ever will, for sharing their knowledge of these hills, and the animals that roam them, with me. So many thanks to Denise Dailey and Virginia Schwartz, my writing companions in the Catskills, both skillful and accomplished authors, who always found time to listen and to encourage my writing of this book—they are true friends. Thanks to Nick Lyons, who knows a lot about the Catskills, publishing and writing, whose encouragement and praise meant the world to me. Special thanks to wildlife biologist Larry Bifaro, who generously and patiently shared his knowledge of black bears in the Catskills with me and who also shared a bear with me, living on the same mountain. Special thanks, too, to Lynn Goldberg, who has always believed in my writing, and thanks especially to Deborah Glick, the best support and critic a writer could have. And the most of all, I thank the critters of the Wild Catskills, the companions, inspiration, and solace of my days.

Index

About the Author

Leslie T. Sharpe is a former Vice President of the New York City Audubon Society and editor of *The Urban Audubon*, an environmentalist, and a naturalist living in the Great Western Catskills. She has taught writing and editing at Columbia and New York Universities. Her book *Editing Fact and Fiction* is a valued resource for authors and editors alike.